MAN, SCIENCE, HUMANISM

MAN, SCIENCE, HUMANISM

A New Synthesis

IVAN FROLOV
Editor of ***Pravda***

PROMETHEUS BOOKS
Buffalo, New York

Published 1990 by Prometheus Books Publishers

With editorial offices located at 700 East Amherst Street, Buffalo, New York 14215, and distribution facilities at 59 John Glenn Drive, Amherst, New York 14228.

Published in cooperation with Progress Publishers

Library of Congress Catalog Number: 90–61869

ISBN 0–87975–626–8

Printed on acid-free paper in the United States of America.

CONTENTS

INTRODUCTION

MAN IN THE AGE OF SCIENCE AND THE SCIENCE OF MAN: PROBLEMS AND OPINIONS

> Man is a mystery that must be
> solved, and if you should spend your
> entire life in solving it do not
> say that you have wasted your time.
>
> *Fyodor Dostoyevsky*

Perhaps never before has the human race experienced such a responsible period so important for its future as the present transition to the third millennium, characterized by a struggle between socialism and capitalism, the threat of a thermonuclear catastrophe, by exceptional manifestations of human genius and by science's penetration into the microcosm, outer space and the deepest foundations of both Nature's life and that of man.

But what does this knowledge yield for man himself? Does it not create new dangers, together with new possibilities for growing wellbeing and for material and spiritual culture? Not surprisingly, the subject of "man and science" has become a major theme in philosophical and sociological studies and is attracting the attention of the world public.

One of the symptoms of the tense concern with which we are approaching the transition to the third millennium is a humanistic, moral evaluation of the current role and the historical mission of science in the life of man and the human race. Is science a force serving the needs and interests of man regardless of the type of society in which it operates and develops, or is its orientation, including possible negative applications, completely determined by society? To what extent can science contribute to man's development or threaten him? What is its correlation not only with the social conditions and factors influencing its development, but also with values, in particular with humanistic principles? Can the latter play a regulating role in that development? How does man change in the course of scientific and technological progress, and how is the latter transformed under the influence of changes in man, his civilization, and his culture?

The significance of these issues is not only theoretical but also largely practical. Yet they, in turn, depend on certain practical needs defined by life and the development of social relations. Naturally, although these questions have many global aspects, they are resolved differently in different social systems. Philosophy, too, plays a role in their theoretical solution, as it addresses itself to their world-view aspects and analyzes the humanistic problems of the present and the future. Today such an examination and analysis possess not only what may be described as constructive scientific and social functions but also a critical function, as they become involved in the ideological and political struggle taking place in the world.

Above all, in cultivating love for wisdom and for man as its carrier and in performing its integrating (methodological) and value-regulating (axiological) functions in modern science, scientific philosophy studies above all *the essence of man,* the general principles that govern his emergence and development, his *goals and ideals,* and the ways that lead to them. It clarifies the mediation and transformation of the natural-biological by the social, the unity of man's essence and existence, the dialectics of man's individual and historical development as the assimilation and reproduction of mankind's socio-cultural experience, man's absolute movement of becoming, and the development of the richness of human nature as an end in itself (Karl Marx).

Scientific philosophy is thus an "integrator", of sorts, of knowledge about man. It correlates this knowledge with general social objectives of and long-term outlook for mankind's development. This particular function of philosophy is especially significant today.

Scientific philosophy observes that at present approaches and methods in studying man are still qualitatively heterogeneous. Unfortunately, these methods are often "complementary", as Niels Bohr would have said, that is, they are mutually exclusive in concrete research. This fact largely explains the level at which the cognition of man still finds itself. A genuinely integrated approach is missing, and man therefore remains "a divided object" of cognition, concerning which we can know everything except what constitutes his integral unity as a biosocial being governed by numerous laws and "systemic forces" emerging in the interaction of many biological, psychological and social factors.

I think that a negative role is played in this respect not only by biologistic approaches but also by individual attempts to portray man as a "social embodiment", thus ignoring the interaction between social and biological factors in man's emergence and development. This is why an integrated approach to the problem of man presupposes a special role for scientific philosophy in analyzing man's essence and existence, anthroposociogenesis and the history of the human race, the development of its civilization in connection with the revolution in science and technology, man's biology and genetics, studies of his psychophysiological abilities and ways of using them in education, the development of man's creative potential, and, finally, the infinite world of his moral life and the humanistic principles of the entire system of human activities, of culture and of the sciences of man.

One may define the subject of *scientific philosophy of man* or of the philosophical science (teaching) concerning man as the *dialectics of human development.* That dialectics, an inherent part of dialectical and historical materialism, permeates all other parts and branches of scientific philosophy as well as its world view and methodological basis, including the theory of dialectics, since all of these are considered in terms of the basic characteristics of *man's relation to the world.*

The above aspect determines a special position and role of scientific philosophy of man not only within the system of concrete sciences studying man from different points of view but also as an integral whole that is frequently associated with the idea of a unified science of man. The latter appears today more as a "regulatory objective", intended to *scientifically* organize and direct integrated studies of man. The above refers, accordingly, not to some currently perceived practical *organization* of that science—the very idea of which, as experience shows, can only produce endless abstract scholastic discussions far removed from concrete recommendations or action—but to the best way to assure, at least theoretically, integrated studies of man *within the existing structure of science.*

Still, I think that the above should not compel us to fully reject the idea of an integral science of man as a *realistic ob-*

jective. The case in point is, obviously, not so much the present practice of scientific cognition as its current development tendency. The appearance of a science of man will conclusively overcome the dualism that still exists today between natural-scientific and sociological methods; thus, both man's biological nature and his social essence will be taken into account. The integral science of man will therefore be a synthesis of many specialized sciences, both natural and social, that study man from different angles. This was foreseen by Marx, who believed that science's ideal future state would be one in which "natural science will ... incorporate into itself the science of man, just as the science of man will incorporate into itself natural science: there will be *one* science".[1] In that connection, Marx observed, "*man* is the immediate object of natural science... But *nature* is the immediate object of the *science of man*."[2] He viewed the terms "human natural science" and "the natural science of man" as identical.

One of the most important tasks of the integral science of man will be to analyze man's future prospects. The global and complex nature of such a task already today makes it a relevant problem of the entire system of concrete sciences of man, including the scientific philosophy of man. It is perhaps in this context that the latter would most convincingly demonstrate that it is authentic, humane and relevant to all of mankind.

Thus an opportunity is also created to raise the question of the very meaning of human existence, and of possibilities for prolonging the duration of life, as well as of death and human immortality, not only in biological terms but also in socio-ethical and humanistic terms. Are the philosophical evaluations of these problems changing, and if so, in what direction, and how will this influence man's prospects in connection with his venture into outer space and search for extraterrestrial civilizations? How will all this influence individual self-awareness today and in the future? Answers to these questions can be found only from a position of new (real) humanism, and in a scientific *philosophy of man, his life, death, and immortality*. In

[1] Karl Marx, "Economic and Philosophic Manuscripts of 1844", in: Karl Marx, Frederick Engels, *Collected* Works, Vol. 3, Progress Publishers, Moscow, 1975, p. 304.

[2] *Ibid.*

other words, with all the importance of *direct* studies of the problems of man's life and death (of projects for man's future biological restructuring, for example) we are not fully solving the basic question, namely, *for what reason* all this is needed if we only consider man's natural-biological existence. Since man as a social being extends beyond the framework of his biological existence, a resolution of that basic question must also extend beyond the sphere of direct studies of man's life and death in the biological sense and turn to philosophy, where the biological and social meanings of life and death of man and of his future as a biosocial being find their true definition as a *dialectical integral unity.*

Having established this unity, it is apparently also possible to outline the basic contours of the "eternal" question concerning the meaning of human life, without an answer to which all other questions relating to man, his natural-biological and social existence and his future become meaningless. The former question has traditionally been addressed by philosophy since the most ancient times. Modern science has achieved much in understanding the biological meaning of human life and death. It has contributed something new to the cognition of their social meaning, without which their specifically human characteristics vanish and only the primeval mystery remains. Outside these poles, life and death, it is impossible to understand the issues currently associated with the problem of longevity, that is, with increasing the duration of human life.

In that connection an important role is played by the socio-ethical, moral-humanistic, and moral-philosophical problems of the cognition of man. In a certain sense the problems that arise in this regard lie outside concrete time: these are "eternal" problems. Another specific feature of these problems is that they are not amenable to unique solutions that would be appropriate for all times. Nor is there a unidirectional movement of cognition whereby each subsequent solution would complement the preceding ones, as it were. This is why in many cases we may find that the moral-humanistic reflections of Socrates or Kant, for example, are more relevant today than many maxims of modern thinkers.

As in the case of many other cultural phenomena that are distinct from science, the specific meaning of moral-humanistic problems is that they involve a permanent dialogue that presupposes a stimulation of active spiritual activity on the part of the subject who experiences what would appear to be unique situations. Nevertheless, an orientation on objective values in this sphere *is* possible, and the search for them reveals certain governing principles that, however, are not inflexible in nature. In particular, these include the development of criteria of the admissibility or inadmissibility, in terms of moral-ethical and humanistic considerations, of particular biomedical experiments, the ethical principles of genetic control, of genetic engineering, of a number of psychophysiological studies, including psychosurgery, in short, experiments that carry a threat to man's health and habitat that are fraught with the danger of personality manipulation and that jeopardize man's inalienable freedoms and rights.

As it turns to man's moral world, the scientific philosophy of man states that there is a dialectical interdependence between the natural-biological and the social, between the finite and the infinite, between man's death and immortality, the complete forms of which, namely, man's material and spiritual culture, alone correspond to his essence. In that sense it establishes certain ideal objectives, which, while creating the feeling of an inseparable integral unity of life, death and immortality, derive, as it were, from the infinity and eternity of the life of the inimitable and unique *personality*. This is, in fact, the basis for the regulating role of morals both in man's everyday life and in his relation to death and immortality.

But that question also involves a different, practical scientific aspect associated with the real, natural-biological infiniteness of the *human race* and the immortality of its reason and humaneness, as well as with the question of whether other forms of life and reason, and other, extraterrestrial, civilizations could exist in the Universe. This raises complex philosophical problems, that also require analysis and discussion from the positions of a new (real) humanism and of scientific philosophy of man. That analysis establishes the uniqueness of *homo sapiens et humanus,* his specificity and emphasizes the global and cosmic destiny of man and the human race.

Such is the diverse spectrum of philosophic problems that emerge as we turn to man in his relation to the development of science, and to the humanistic aspects of these problems. Today attempts to separate and juxtapose science and man, research-oriented and value-oriented humanistic approaches, appear to be receding into the past. The tendency to determine the "human dimension" of science and of scientific and technological progress is becoming increasingly evident, and the world scientific community is becoming increasingly aware of the need for it.

Occasionally, however, attempts to somehow combine man and science in a humanistic synthesis have an ambiguous character. Frequently they are not substantiated in terms of scientific philosophy and do not rest on reliable practical foundations. This is why the task of arriving at a *new synthesis* of man, science and humanism remains relevant. In solving it we turn to Marxism, for it is precisely along that path, I believe, that the new synthesis, concerning which so much has been written from positions of different world views, can be found.

The task of this book is to show all this in a concrete analysis of concrete humanistic problems of man in the age of science and scientific and technological progress, and in the context of practical problems the solution of which ushers in "the age of man". At the same time, the author is aware that, having subjected individual approaches to a critical analysis and rejected false views, we should keep it in mind that we shall not have attained the "ultimate truth". In seeking to solve a specific problem we must take a fully critical approach to our own generalizations and conclusions, and must view them merely as material for further reflections and for the development of objectively true positions with regard to the new vital questions that concern us. A general formulation of many problems of man in his relation to science may produce a variety of concrete approaches to their further discussion, including those within the framework of Marxist-Leninist philosophy. Nor is this unexpected, if one considers the novel character of and the insufficiency of previous research into these problems, the open character of Marxist philosophy, and its capacity for development and renewal under the influence of progress in science and in social reality.

Still, we must proceed from ideals and principles that have been proven by scientific philosophy and historically. Let us consider what these ideals and principles are, and how scientific philosophy has embraced them in the course of a painful and contradictory search, scientific polemics and struggles, including the struggle against today's numerous myths and pseudo-scientific theories.

The entire history of human cognition has been characterized by an endless search for human essence and attempts to unravel man's mystery. The latter, however, still remains in many ways a mystery, even though science now relies on highly complex methods as well as the historical experience of many centuries.

At the same time, decisive landmarks have been reached in the history of cognition of man's essence, and these make it possible to gradually deepen and widen our search at new stages in the development of science, of mankind's spiritual culture and of its social experience. These landmarks have been reached thanks above all to the genius of Marx. His teaching concerning man is closely associated with a materialistic understanding of history and a recognition of the decisive role that is played by labour and by the practical activities of individuals in the life of society, with the understanding of the social essence of the individual, and with the conception of scientific, real humanism. Marxism also provides an answer to the question concerning the correlation of the social and the natural-biological aspects in man, and whether it is the social or the natural-biological factors that play a decisive role in man's development.

If one takes a retrospective view of the history of philosophy one will see that the problem of man has been generally approached with the help of methods that were somehow "complementary", i.e., with idealistic methods that assigned an absolute character to man's spiritual origins, and naturalistic methods whereby the primary importance was attached to man's natural-biological features. That dualism of the soul vs. the body has been an eternal theme of philosophical interpretations of man, from ancient times until the present. At the same time, in the history of philosophy the problem of man was also viewed as a question of internal vs. external factors determining the nature of man. By external factors, that is, the environment within which man's formation and development take place, philosophers naturally meant not only natural factors but also social ones, such as education.

There is no need in the present context to go back into the history of attempts to find the essence of man: while this is an extremely absorbing task, it would lead us away from the problems that we are considering.[1] I would only like to stress that in its understanding of the essence of man Marxism proceeds from the historical tradition and inherits everything that is valuable in it, from antiquity to the present. At the same time, Marxism has created a fundamentally new understanding of the essence of man, and credit for this goes above all to Marx himself.

While Marx's sociological theory may appear ancient history, in fact the situation is quite different: the laws that were discovered by Newton, for example, or by Darwin, have not receded into the past; they also operate in our time, even though many new discoveries have been made that develop and complement them. This is also true of the discovery of the essence of man that was made by Marx in the last century: as I will seek to show, it has lost none of its significance in the light of the findings of modern science. What is more, Marx's discovery acquires even greater significance in the light of these findings, although its importance is either played down or else distorted by opponents of Marxism, and thus initially requires an adequate presentation. This is why numerous quotations will now follow, for it is best to listen to Marx himself rather than to those who interpret him.

In contradistinction to the idealistic, religious-mystical conceptions and naturalistic anthropologism that prevailed before Marx in the understanding of man's essence, Marx formulated a proposition that has become a key element in the scientific understanding of man: "The essence of man," as he defined it, "is no abstraction inherent in each single individual. In its reality it is the ensemble of the social relations."[2] To Marx it was clear from the outset that the essence of man is not "its abstract physical character, but its *social quality*"[3] and that, accordingly,

[1] That is the subject of my book *Perspektivy cheloveka* (Prospects of Man), Politizdat, Moscow, 1979, 1983.

[2] Karl Marx, "Theses on Feuerbach", in: Karl Marx, Frederick Engels, *Collected Works*, Vol. 5, 1976, p. 4.

[3] Karl Marx, "Contribution to the Critique of Hegel's Philosophy of Law", in: Karl Marx, Frederick Engels, *Collected Works*, Vol. 3, p. 21.

"all history is nothing but a continuous transformation of human nature".[1] In his view, however, we "must first deal with human nature in general, and then with human nature as modified in each historical epoch".[2]

Marx sought a definition of man's essence in the specifically human activities and existence and found it in labour. It was labour that created man and it is in labour that specifically human qualities are reflected and developed. In the study entitled *A Contribution to the Critique of Political Economy,* Marx defined human labour as "useful activity directed to the appropriation of natural factors... labour is a natural condition of human existence, a condition of material interchange between man and nature, quite independent of the form of society."[3] Labour, to Marx, is not only a means of existence but also a form of man's self-assertion as in labour man "really proves himself to be a *species-being*".[4]

In this context Marx viewed labour above all as a social relation towards Nature; consequently, man's social character is of necessity included in the definition of man's essence. Man is inseparable from society, for only in society, in specific social formations he is realized as man.

Marx and Engels stressed that a "mode of production must not be considered simply as being the reproduction of the physical existence of the individuals. Rather it is a definite form of activity of these individuals, a definite form of expressing their life, a definite *mode of life* on their part."[5] It is for the above reason, in particular, that "the history of *industry* and the established *objective* existence of industry are the *open* book of *man's essential powers,* the perceptibly existing human *psy-*

---

[1] Karl Marx, "The Poverty of Philosophy", in: Karl Marx, Frederick Engels, *Collected Works,* Vol. 6, 1976, p. 192.

[2] Karl Marx, *Capital,* Vol. I, Progress Publishers, Moscow, 1974, p. 571.

[3] Karl Marx, *A Contribution to the Critique of Political Economy,* Progress Publishers, Moscow, 1977, p. 36.

[4] Karl Marx, "Economic and Philosophic Manuscripts of 1844", *op. cit.,* p. 277.

[5] Karl Marx and Frederick Engels, "The German Ideology", in: Karl Marx, Frederick Engels, *Collected Works,* Vol. 5, p. 31.

chology. Hitherto this was not conceived in its connection with man's *essential being*."[1]

Man, Marx emphasized, is the "permanent pre-condition of human history, likewise its permanent product and result".[2] The development of the richness of human nature, consequently, involves the development of man as a personality in the process of his material and spiritual activities, interaction with other persons, in learning, training, and education as an assimilation and reproduction of mankind's socio-cultural experience, progress in social relations and way of life, of consciousness and self-awareness, and of moral self-perfectioning.

At the same time, Marx emphasized that personality formation is a complex and contradictory process and that the manifestations of its individuality are ambivalent. He drew attention to the distinction between the individual as a personality and the accidental individual who is incapable of realizing himself as a personality, of developing his potential and abilities through activity and a way of life corresponding to his essence, that is, truly human ones. Such a development of man is possible only under socialism and communism, which "ensures, together with the greatest expansion of the productive powers of social labour, the most complete development of man".[3] Communism can create "for *all people* such a condition that everyone can freely develop his human nature and live in a human relationship with his neighbours".[4] This is why communism is "the age of man", a new civilization creating a new man both as a result of social development and as its condition and moving force, i.e., as its "end in itself".

Such is the position of Marxism on the *social essence* of man and the historical forms of its realization in the course of social development. Theoretical analysis has produced a diver-

---

[1] Karl Marx, "Economic and Philosophic Manuscripts of 1844", *op. cit.*, p. 302.

[2] Karl Marx, *Theories of Surplus-Value*, Part III, Progress Publishers, Moscow, 1975, p. 491.

[3] "Marx to the Editorial Board of the *Otechestvenniye Zapiski*", in: Marx, Engels, *Selected Correspondence*, Progress Publishers, Moscow, 1975, p. 293.

[4] Frederick Engels, "Speeches in Elberfeld", in: Karl Marx, Frederick Engels, *Collected Works*, Vol. 4, 1975, p. 263.

sity of grounds for its logical identification in historical studies of anthropogenesis, the *formation* of the social essence of man, and, more generally, of his development as a species-being. Here again Marxism continues to exert its effective heuristic theory of anthropogenesis. In his *The Descent of Man, and Selection in Relation to Sex* (1871) and *The Expression of Emotions* (1872) Charles Darwin, the founder of the theory, initiated the development of scientific conceptions concerning man's origins and the human race's historical formation. This followed logically from his general conception of evolution. Already in the first edition of "Origin of Species" (1859), Darwin noted that "by this work 'light would be thrown on the origin of man and his history;' and this implies that man must be included with other organic beings in any general conclusion respecting his manner of appearance on this earth."[1] Darwin substantiated that conclusion by examining the descent of man from a certain primeval type and by following his mode of development and determining the significance of differences among human races.

Darwin approached this issue above all as a problem in *natural science,* as opposed to the religious dogma of the "divine creation" of man. It is precisely from this angle that he examined the process of anthropogenesis, its origins and motive forces. Having convincingly demonstrated man's natural origins, Darwin also laid the groundwork for a subsequent scientific explanation of the *social* mechanisms relating to the formation of the human race, associated with the emergence of labour, consciousness, speech, and various modes of social interaction. In that explanation a decisive role was played by his methodological approach, that followed directly from a Marxist understanding of man's essence as the ensemble of all social relations and representing a summing-up "of the most general results, abstractions which are derived from the observation of the historical development of men".[2]

Thus Marxism extended a historical approach, in which general evolutionary and sociological ideas and methods are dia-

[1] Charles Darwin, *The Descent of Man, and Selection in Relation to Sex,* John Murray, London, 1890, p. 1.

[2] Karl Marx and Frederick Engels, "The German Ideology", *op. cit.,* p. 37.

lectically combined, to the origins and formation of man and the formation of society, combining them into an integral whole. This was done in a systematic form by Engels in a study entitled "The Part Played by Labour in the Transition from Ape to Man" (1876). But the position of principle on this issue was formulated by Marx and Engels much earlier, namely, in preparatory studies for *The Holy Family* (1845), in *The German Ideology* (1845-1846), in *A Contribution to the Critique of Political Economy* (1859), and in the first volume of *Capital* (1867). In stressing that a transition from animals to man constituted the emergence of a new quality,[1] Engels observed that "the essential difference between human and animal society consists in the fact that animals at most *collect* while men *produce*. This sole but cardinal difference alone precludes the simple transfer of laws of animal societies to human societies."[2]

The role of these ideas as fundamental principles of a scientific understanding of anthropogenesis as it relates to sociogenesis was fully confirmed by the subsequent history of anthropology. Soviet anthropologists have substantially developed the Marxist historical approach, even though there have been differences of opinion among them concerning many concrete questions of anthropogenesis and occasionally highly contradictory views have been expressed. Moreover, anthropological studies proper, which trace the formation of man's social essence, have been combined with a special analysis of the emergence and development of social production. In some studies, one even finds formulations of a synthetic conception of anthroposociogenesis.[3] Its essence is that natural biological selection was gradually replaced in the course of anthropogenesis by sociobiological or biosocial selection. One can, of course, argue about concrete

---

[1] See: "Engels to Marx in London, Manchester, July 14, 1858", in: Marx, Engels, *Selected Correspondence,* p. 102.

[2] "Engels to P. L. Lavrov in London, November 12-17, 1875", in: Marx, Engels, *Selected Correspondence,* p. 284.

[3] See: Yu. I. Semenov, *Kak vozniklo chelovechestvo* (How Mankind Emerged), Nauka, Moscow, 1966; *idem., Proiskhozhdeniye braka i semyi* (The Origin of Marriage and the Family), Mysl, Moscow, 1974; Yu. I. Yefimov, *Filosofskiye problemy teorii antroposotsiogeneza* (Philosophical Problems of the Theory of Anthroposociogenesis), Nauka, Leningrad, 1981.

details of that hypothesis, but one thing is clear, namely that "the social determinant" in the process of anthropogenesis, or the combining of anthropogenesis with sociogenesis is a necessary element of a Marxist historical approach to man's origins and development.

It is true that if we limited the scientific definition of man, given by Marx, to merely stating that his essence is social, the resulting "social embodiment" would be a far cry from the real, concrete man. But such an approach was alien to Marx, who viewed the history of the human race as the outcome of the activities of full-blooded, living, thinking and feeling man, endowed not only with reason but also with feelings, and experiencing sometimes painful contradictions between the commands of reason and those of his "nature", his passions and suppressed instincts, between consciousness and "the subconscious", no matter how we call this. Even in his most complex and highly abstract theoretical expositions (for instance, in *Capital,* in which the force of abstraction was the major means of cognition) Marx constantly focused on the real man.

Let us recall the vivid and colourful character of many historical personalities in the works of Marx. In criticizing the bare abstractions concerning some subjectless "history" of the human race Marx and Engels wrote that "history is *nothing but* the activity of man pursuing his aims".[1]

Opponents of Marxism continue to refer to some kind of "super-rationality" characterizing Marxism in its understanding of history and of man's role therein which, it is asserted, even if recognized by Marxism, is seen only as a certain deduction from general social premises. This approach fully ignores what Marx described as "man's nature". The distinction between man's essence as a personality and his existence as an individual and as a representative of the human race is thus ignored. Accordingly, a more comprehensive definition would note the following: *man is as a subject of the socio-historical process of development on Earth of material and spiritual culture, a bio-social being possessing articulate speech and consciousness, genetically linked with other forms of life, from which, however,*

[1] Karl Marx and Frederick Engels, "The Holy Family", in: Karl Marx, Frederick Engels, *Collected Works,* Vol. 4, p. 93.

he has evolved through his ability to produce instruments of labour.

In Marxism the definition is elaborated to the point of distinguishing man's social essence as the ensemble of all social relations, that is not, however, juxtaposed to his existence as a natural-biological being. At the same time, social factors do determine natural-biological ones to a substantial extent. What is important for us, however, is to examine them separately in order to show that scientific Marxist-Leninist philosophy of man, the origins and development of which are closely associated and interact with the entire system of the sciences of man, establishes man's essence not by separating it from the natural-biological factors of his existence but in an interaction and interrelationship with them. This makes it possible to overcome both the extremes of vulgar sociologism and those of social-biologism, and to define dialectically the direction of the search for the prospect of man as a biosocial being in terms of both his individual and historical development.

It must be pointed out that the conception of man as a biosocial being merely notes specific features of his actual, bodily and intellectual, existence, namely, the fact that he belongs simultaneously both to the natural-biological and the social world. Of course, externally defined features of human existence cannot be extended to encompass his internal characteristics, since this may lead to false conceptions concerning man's "biosocial essence", etc. While man's essence is entirely social, his existence is not defined by this factor alone but is in many ways associated with natural-biological factors.

The scientific approach to this question is quite comprehensively described in the works of numerous Marxist authors. It is true that there are differences in the interpretation of particular aspects of the overall problem, but these differences need hardly be overemphasized to the point of implying differences in world views or ideologies.

Our task is to establish more deeply and more comprehensively the dialectics of the social and the biological in man. This does not mean that we simply note a specific correspondence between them that is given once and for all, even if we do refer to the primacy and predominance of the social in relation to the biological. Rather, the dialectics consists in the mediation and transformation of the biological by the social, since "*human* objects are not natural objects as they immediately present themselves".[1] This is also true of man's biological needs; the nature of their transformation is deep enough for Marx to have observed that "hunger is hunger; but the hunger that is satisfied by cooked meat eaten with knife and fork differs from hunger that devours raw meat with the help of hands, nails and teeth."[2]

The basis and the mechanism of the mediation and transformation of natural-biological factors by social ones are inherent in specifically human activities, above all labour and production, which create biosocial forms of man's existence and development.

In recent years, the problem of the mediation and transformation of natural-biological elements by social ones in the process and as a result of human activity has become the subject of in-depth and increasingly comprehensive study in Marxist science. It has been shown that individual development is determined socially; the foundations of a philosophic-sociological conception of man's ontogenesis have been developed, as have conceptions of the latter's integral character and stage-by-stage development; the socially determined nature of the biological factors in man, including his genetics, ontogenetic and general physical development, adaptation, and ecology, has been given a consistent dialectical-materialistic interpretation. In anthropology the view is developing that anthropogenesis does not simply result from a dominant influence of socio-historical factors, but that at a particular stage it fully overcomes the forming influence of biological laws. The dominant influence of social factors in the historical process has been shown. In psychology there is an increasing acceptance of theories that develop the views of Lev Vygotsky concerning the leading role of social factors in the formation of man's psychology. In genetics one currently finds an increasingly definite emphasis on the mediating role of social factors in relation to biological ones and on the mechanisms that transform the latter in man's behaviour.

---

[1] Karl Marx, "Economic and Philosophic Manuscripts of 1844", *op. cit.*, p. 337.

[2] Karl Marx, *A Contribution to the Critique of Political Economy*, p. 197.

It has already been noted that different authors naturally view many aspects of the problem of the social vs. the biological in man in different ways. And it is not surprising that scientific discussions have been taking place among Marxist scientists in recent years concerning this complex and apparently insufficiently studied problem. I believe that these discussions largely account for the specific way in which the problems of the social vs. the biological are formulated in science today. At the same time, from different points of view, the element which is fundamental in a scientific resolution of the problem, namely, the mutual interaction and mutual influencing of social and biological factors in man, whose social essence is by no means contrasted to his biological nature, is stressed.

Today this general dialectical-materialistic solution of the problem finds itself in opposition to numerous attempts to develop methodological approaches based on idealistic anthropologism and naturalism that have been scientifically overcome in the Marxist teaching concerning man, and therefore often appear as "regressions of the regressions" that have emerged after Marx's discovery of man's social essence and of his development of a scientific conception of anthropogenesis and prospects for man's historical development. These include philosophical-anthropological, social-Darwinistic, Freudian, and similar concepts. This does not mean, of course, that nothing new has been created in their philosophic and especially their concrete-scientific substantiations in the context of modern conditions. In particular, in connection with the intense progress of biological sciences naturalistic conceptions of man that assign an absolute role to his biological nature have been developing particularly intensively in various trends in modern social-biologism.

For example, in opposition to the scientific conception of human races that has been developing in ethnic anthropology, or the study of races, "racial theories" have begun to appear that continue to nourish racism, one of the most repulsive products of bourgeois thought, that has found its most extreme expression in the fascist practice of genocide. These "theories" were first formulated by Joseph de Gobineau, the author of *Essai sur l'Inégalité des Races Humaines* (1853-1855) which all types of racists, including the so-called "new right" in France, continue to advocate to this day, even though science, beginning with Darwinism and up to modern biology and genetics, completely rejects racist fabrications concerning the inequality of human races by demonstrating the unity of the human races based on their belonging to a common species.[1]

This is also true of the so-called social Darwinism which displays an inherent kinship to racism.[2] That trend, which draws its origins from Herbert Spencer, has nothing in common with genuine Darwinism, even though it does borrow from it some terminology. Such "theoreticians" of social Darwinism and of racism as Ludwig Gumplovicz, Gustav Ratzenhofer, Ludwig Woltmann and Houston Chamberlain have gone to extreme lengths in biologizing man and society in a belief that it is only through natural selection that it would be possible to avoid a "genetic" degeneration of the human race. In this way they established certain foundations for modern social-biologism, which seeks to associate itself with the prestige of the newest findings in genetics, ethology, and other sciences.

Freudism is still another source of modern naturalistic conceptions of man based on a sociobiological approach. One could observe a clear identification of man's essence with biologically established asocial instincts already in the works of Sigmund Freud, the founder of modern conceptions of psychoanalysis. According to Freud, social factors only play a role in suppressing and supplanting biological instincts. From this he concluded that Eros, the sexual instinct of life and self-preservation, and Thanatos—the destructive and aggressive instinct of death—constitute the basis of social progress. The levels of culture in man's development were graded accordingly. It should be noted that Freud drew the attention of scientists to the sphere of the unconscious, which traditional branches of science had not studied systematically, and pointed to the great importance of that sphere in human life and in cultural development. But he not only greatly overemphasized its role but was generally unable to uncover the meaning that man's natural-biological strivings

---

[1] See: *Rasy i obshchestvo* (Races and Society), Nauka, Moscow, 1982.

[2] In *Die Zerstörung der Vernunft,* (Aufbau-Verlag, Berlin, 1955), G. Lukács, the Hungarian Marxist philosopher, provided a criticism of racism and social Darwinism in their historical aspect and in their connection with fascism.

acquire within the social sphere of his existence. Biologistic tendencies are especially evident in modern Freudism.[1]

It should be noted that the natural-biological determinant is greatly overemphasized in many trends in bourgeois philosophical anthropology. This originated in the works of Friedrich Nietzsche and Max Scheler, but it is in modern versions of philosophical anthropology, including existential ones, that it is most developed. Nietzsche's views were based on a general conception that man's nature is impervious to consciousness and reason and is subordinated instead to an unconscious creative life and pressures of a chaotic abundance of instincts. Nietzsche viewed man as "a still unformed animal" that has not yet adapted itself to its natural-biological existence and is accordingly open to all other possibilities. Max Scheler, in his study entitled *Die Stellung des Menschen im Kosmos*[2] and Helmuth Plessner, in *Die Stufen des Organischen und der Mensch*[3], developed further these ideas of Nietzsche and combined them with Freudian conceptions concerning the unconscious and its role in man's life and activities. It is such a biological-anthropological approach that has been a specific characteristic of the philosophic approach to the understanding of man, which is clearly evident in its subsequent modifications.

This "new philosophical anthropology", as Scheler himself called it, claimed the status of a philosophical theory of a scientific knowledge of man, capable of providing answers to all fundamental questions of human existence, including man's essence, the natural-biological, psychological and spiritual factors in his existence, as well as personal and social factors. In seeking to show the "open character" of man through a comparison with animals, Scheler stressed that man's existence (being) results from a tension between life and the spirit which he viewed as a principle in opposition to any form of life and which could

---

[1] See: *Neue Perspektiven der Psychoanalyse.* Ed. by W. Mertens, Kohlhammer, Stuttgart, etc., 1981; B. A. Farrell, *The Standing of Psychoanalysis,* Oxford University Press, Oxford, etc., 1981.

[2] Max Scheler, *Die Stellung des Menschen im Kosmos,* Reichel, Darmstadt, 1928.

[3] Helmuth Plessner, *Die Stufen des Organischen und der Mensch,* Gruyter, Berlin and Leipzig, 1928.

not be reduced to the natural evolution of life, even though it is precisely through sublimation and the suppression of impulses that it acquires its force and power. At that point "the new philosophical anthropology" arrives at the idea of a God that is not yet "perfect" and "complete" but "is in the making", a process in which man occupies a central position.

Scheler's attempt to combine philosophical, natural-scientific (primarily biological) and religious approaches that break with traditional views into "a new philosophical anthropology" was continued in different forms, and with varying emphases on the role of particular properties of "human nature", above all in the philosophical-anthropological constructions of Arnold Gehlen[1], Otto F. Bollnow[2], and others who developed further the biological aspect of that conception.

Cultural-philosophical anthropology seeks to overcome bio-anthropological tendencies, at least in their absolute form. Its main representatives include Erich Rothacker[3] and Michael Landmann[4], who, on the basis of Plessner's work, seek to comprehend man in his integral unity, i.e., in the unity of his natural-biological existence and spiritual activity as both the creator of culture and its creature. In seeking to overcome one-sided biologism, however, cultural-philosophical anthropology places an excessive emphasis on another extreme position. Michael Landmann, for example, denies the role of biological evolutionary factors in understanding man's nature altogether. Such an exclusively "cultural-philosophical" approach is combined with a recognition of an "objective spirit" preceding man, and that already leads to a philosophical-religious version of anthropology that is espoused, in particular, by Hengstenberg.[5]

As a result, one observes in the development of modern bourgeois philosophical anthropology a feature that was also char-

---

[1] Arnold Gehlen, *Der Mensch, seine Natur und seine Stellung in der Welt,* Junker und Dünnhaupt, Berlin, 1944.

[2] O. F. Bollnow, "Die Philosophische Anthropologie und ihre Methodischen Prinzipien", in: *Philosophische Anthropologie Heute,* Munich, 1972.

[3] Erich Rothacker, *Philosophische Antropologie,* Bouvier, Bonn, 1966.

[4] Michael Landmann, *Philosophische Anthropologie,* Gruyter, Berlin, 1955.

[5] Hans-Eduard Hengstenberg, *Philosophische Anthropologie,* W. Kohlhammer Verlag, Stuttgart, 1957.

acteristic in the past, namely, modern anthropologism and naturalism are both based methodologically on a juxtaposition of integral and reductionist methods in the cognition of man. It is true that in both cases there is a certain arbitrariness in the juxtaposition, since an anthropological approach may be accompanied by a naturalistic reduction of man, in particular, by viewing his biological nature in isolation from other dimensions. Reductionist tendencies, on the other hand, may require an assertion of integral, and above all philosophical, approaches to the problem of man.

Soviet Marxists have given a sufficiently comprehensive and persuasive consideration to the basic variants of modern philosophical anthropology.[1] In particular, it was shown that in spite of a number of substantial differences they possess common historical origins and a common methodology. With all the differences among different variants of modern philosophical anthropology, they are similar in their main aspect, namely, a propensity to view the problem of man as exclusively (or primarily) the task of philosophy. As a result, the gap between philosophy and concrete sciences concerned with man is made even deeper, even though claims are often made that philosophical anthropology is closely associated with concrete sciences and makes constant use of their findings[2]. In particular, this is characteristic of the biological and psychological variants of philosophical anthropology which only appear to rely on corresponding sciences

---

[1] See: *Problema cheloveka v sovremennoi filosofii* (The Problem of Man in Contemporary Philosophy), Nauka, Moscow, 1969; *Burzhuaznaya filosofiya XX veka* (Bourgeois Philosophy of the Twentieth Century), Politizdat, Moscow, 1974; R. M. Gabitova, *Chelovek i obshchestvo v nemetskom ekzistentsializme* (Man and Society in German Existentialism), Nauka, Moscow, 1974; G. M. Tavrizyan, *Problema cheloveka vo frantsuzskom ekzistentsializme. Kriticheskij analiz* (The Problem of Man in French Existentialism. Critical Analysis), Nauka, Moscow, 1977; *Lichnost v XX stoletii. Analiz burzhuaznykh teorii* (The Personality in the Twentieth Century. An Analysis of Bourgeois Theories), Mysl, Moscow, 1979; P. V. Korneyev, *Kritika sovremennykh burzhuaznykh i revizionistskikh kontseptsii cheloveka* (A Critique of Contemporary Bourgeois and Revisionist Conceptions of Man), Vysshaya shkola, Moscow, 1981; B. T. Grigoryan, *Filosofskaya antropologiya. Kriticheskij ocherk.* (Philosophical Anthropology. A Critical Outline), Mysl, Moscow, 1982, and others.

[2] See: Walther Bruning, "Philosophische Anthropologie", in: *Die Philosophie im XX Jahrhundert,* Ed. by F. Heinemann, Ernst Klett Verlag, Stuttgart, 963, pp. 561-577.

but in fact are as remote from them as are the existential variants, in which (as, for example, in the teaching of Jean-Paul Sartre) a resolute break with science is a basic principle. Such an "ontology of man" assigns a central position to the concepts of subjectivity and transcendentality, continuing self-awareness and self-finding, and freedom in a sense that is isolated from objective principles governing society.

As a result, even in those variants of philosophical anthropology that refer to the social, one finds a reduction of the social to the individual, i.e., a neglect of man's actual social essence as the totality of all social relations. This may be clearly seen, for example, in the existential conception of Karl Jaspers. He believes that, in general, the human personality exists without any connection with history which influences the personality only outwardly; what dominates it is the absolute, eternal, timeless, non-historical. According to Jaspers the social element constitutes man only as an abstract species essence, as an individual. Such an abstract philosophical approach does not, of course, make any clearer the question of the correlation of social and natural-biological factors in human development.

We have already seen that the other extreme position in answering that question is represented by the biological variety of philosophical anthropology and Freudism. Of some interest in that connection are the philosophical-anthropological views of Teilhard de Chardin, who seeks to find mystical linkages between the operation of natural-biological factors in the formation of the human species and social factors. In the case of Teilhard de Chardin the experimental approach of a natural scientist coexists with the mystical faith of a theologian, and this permeates his overall philosophical anthropology. On the one hand, he offers a biologistic conception of man in which, in effect, there is nothing social, in the sense of qualitatively suprabiological; man is seen only as "the arrow pointing the way to the final unification of the world in terms of life. Man alone constitutes the last-born, the freshest, the most complicated, the most subtle of all the successive layers of life."[1] Accordingly, in Teilhard's thinking, a transition to "super-life" is essentially a

---

[1] Pierre Teilhard de Chardin, *The Phenomenon of Man,* Fontana Books, Collins, 1955, p. 247.

biological revolution leading to a conscious transformation of man's psychology, biology and genetics. In this connection he is concerned that "we have given too little thought to the question of what medical and moral factors must replace the crude forces of natural selection should we suppress them",[1] and concludes that in the course of the coming centuries a nobly human form of eugenics is indispensable.[2]

According to Teilhard man's development is proceeding in the direction of "human energetics" that is created not only with the help of but also "beyond all physics, all biology and all psychology... It is in the course of that creation, already obscurely begun, that science, by being led to concentrate on man, will find itself increasingly face to face with religion."[3] This leads Teilhard to oppose a scientific materialistic interpretation of man, although it is true that by the latter he has in mind primitive and vulgar forms of materialism. In his opinion it is through love and through laws of social heredity, which he identifies with the psychological, oral or written transfer of human experience, that man ascends towards a "super-life" and "super-personality" (i.e., towards God).

This inherently contradictory character of Teilhard's philosophical anthropology shows that it is not always possible to clearly define the polarization between "pure" anthropologism, on the one hand, and naturalism and biologism, on the other, for between these two extreme poles there exists a theoretical space that is occupied by "transitional" forms. Nevertheless, in the current context one generally notes a clear tendency to move away from "pure" anthropologism, whose position, still, is quite strong and in a number of cases is gaining further strength due to certain processes in the revolution in science and technology. Instead, moving increasingly to the foreground are the conceptions of social-biologism, which includes heterogeneous theories based on common methodological foundations.

The increasing emphasis in recent years on conceptions of social-biologism that universalize man's biological nature, as opposed to his social essence, may be attributed, in terms of

---

[1] *Ibid.*, p. 310.
[2] *Ibid.*
[3] *Ibid.*, p. 311.

gnosiology, to one-sided and frequently distorted interpretations of findings of modern neurophysiology, etology, and genetics. In that connection it has been asserted that man is a mistake of evolution and that nature has endowed him with a number of negative genetically conditioned attributes constituting a kind of delayed action bomb that man can defuse only if he understands its timing device (Arthur Koestler). A number of etologists believe that certain specific characteristics of human behaviour, such as man's assertedly inborn aggressiveness or, alternatively, his altruism, are also genetically determined.

Biologistic tendencies reveal their scientific inconsistency and disorienting socio-ethical role in the conceptions of modern social Darwinism, although it is true that having become widespread towards the end of the last century and in the early part of the present century, the latter now represents just one approach within other theories and conceptions. In particular, the basic premise of social Darwinism concerning the presence of a struggle for survival in human society, natural selection, etc., is now defended not only in etological but also in certain anthropological-genetic conceptions. This is convincingly illustrated by the widely commented study of Jacques Monod, a Nobel Prize winner, entitled *Le Hasard et la Nécessité*[1] in which he totally biologizes the problems that are associated with human development. At the same time, left out of his analysis are social factors determining human essence, including possibilities for man's further physical improvement, the development of culture, and also the production activities of individuals as the basis of social progress.

Theodosius Dobzhansky, a prominent American geneticist, has also shown a certain overemphasis on the importance of biological mechanisms in human life and society. He believed that natural selection not only continues to operate in modern society but will continue to do so in the foreseeable future.[2] In spite of this his works cannot be associated with the general approach that is described here as social-biologism, for Dob-

---

[1] Jacques Monod, *Le Hasard et la Nécessité. Essai sur la Philosophie Naturelle de la Biologie Moderne,* Editions du Seuil, Paris, 1970.

[2] Theodosius Dobzhansky, "Natural Selection in Mankind", in: *The Structure of Human Populations.* Ed. by G. A. Harrison and A. J. Boyce, Clarendon Press, Oxford, 1972, pp. 213-233.

zhansky, an outstanding scholar, has considered only individual aspects of the problem of man from a highly specific, namely, genetic point of view.

On the other hand, Ernst Mayr, who is a well-known American biologist studying evolution, interprets these problems in a somewhat different context.[1] While demonstrating the advantages of a population-oriented and historical approach to human biology he does not avoid complex social problems, although the solutions that he proposes are not always convincing from a Marxist point of view. At the same time, Mayr's overall world view and concrete approaches to the problem of man and of his future are based on lofty humanistic principles and a recognition of the uniqueness and freedom of the personality.

Aside from the approaches of modern social-biologism that have been mentioned there is also the wide sphere of modern Freudism, which makes intensive use of findings concerning the role of psychophysiological factors in man's vital activities. This is characteristic, for instance, of the psychoanalytical conceptions of Carl Jung and of Karen Horney, among others. Attempts have also been made to combine Freud's psychoanalysis with the Marxist conception of the personality, as, for example, in the works of Erich Fromm. In seeking to draw closer these two diametrically opposed conceptions Fromm has sought to surpass them, interpreting Marxism in a vulgarized sociological and anthropological spirit. Fromm writes that "the analysis of society and of the historical process must begin with man, not with an abstraction, but with the real, concrete man, in his physiological and psychological qualities. It must begin with a concept of the essence of man, and the study of economics and of society serves only the purpose of understanding how circumstances have crippled man, how he has become alienated from himself and his powers."[2] This statement contains the essence of Fromm's conception, which brings it close to the bioanthropological constructions of Scheler, Gehlen, and others, but with an emphasis on psychophysiological factors in their Freudian interpretation. While Fromm defines his point of view as one that is "neither

[1] See: Ernst Mayr, *Populations, Species, and Evolution,* Harvard University Press, Cambridge, Mass., 1971.

[2] Erich Fromm, *The Sane Society,* Rinebart & Company, New York, Toronto, 1955, p. 254.

'biological' nor 'sociological' ";[1] it differs from Freud's conception in that it replaces the operation of instincts by psychological factors. In itself, the psychoanalytical approach to man remains nearly unchanged, since, according to Fromm, social conditions merely produce specific manifestations of man, while the essence of the latter is conditioned by hidden strivings and conflicts within the unconscious, rather than determined socially. It is precisely these factors that, in Fromm's view, play a decisive role. In his opinion, Marx's "error" was that he considered man to be a reasonable being, while in fact man, as demonstrated by Freud, is allegedly an irrational animal.

All this leads Fromm to utopian constructions of the future, whereby man's relations with science, technology and the natural environment would acquire a harmonious integral unity. In his view, however, this would be achieved not as a result of the rational operation of social factors but as the consequence of an effective inherently experienced "revelation", whose role would be to free man's original potential and block the forces of evil, and to draw man closer to a state of unencumbered natural behaviour. These ideas were developed by Fromm on the basis of his specific understanding of man's essence and promotion of the ideas of a "humanistic socialism".[2] In his later works, moreover, he sought to combine as much as possible a philosophical-anthropological approach with the culturological and socio-political ones, expressing an essentially utopian hope for a "new humanism" and a "new society" that would be reformed in accordance with man's nature.

A similar approach was taken by Herbert Marcuse; like Fromm, he engaged in "correcting" Marx by introducing "the biological dimension" of man that assertedly was absent in Marx. In his attempt to biologize Marxism, moreover, Marcuse proceeds further than does Fromm.[3] Like other representatives of the so-called Frankfort school, however, he does not associate himself with any particular trend in bourgeois philosophical anthropology. As will be shown in the next chapter, these au-

---

[1] Erich Fromm, *op. cit.*, p. 14.

[2] Erich Fromm, *The Revolution of Hope*, Bantam Books, Toronto, 1968.

[3] Herbert Marcuse, *Eros and Civilization: a Philosophical Inquiry into Freud*, Beacon Press, Boston, 1955.

thors relate their views to a wider context of theories and ideas, in particular, with regard to the problems of the current revolution in science and technology.

It should again be stressed that within the highly diverse modifications of modern anthropologism and naturalism the sociobiological determinant is continually gaining in strength. At times it is used for quite divergent purposes. For example, American psychologist Burrhus Skinner uses it to oppose the concept of an "autonomous man" possessing freedom of choice in his behaviour. In his opinion the environment alone programmes human behaviour, and this makes it possible to manipulate man with the help of "behaviourist engineering" and to create an absolutely manageable man within an absolutely manageable society which would thus be able to resolve all its problems.[1] In other approaches man's biological nature, as expressed in his genotype, the unconscious and instincts, is viewed as the decisive factor.[2]

A particular role within the multitude of forms of modern biologism is beginning to be played by sociobiology, a controversial scientific approach that aspires to become a distinct scientific discipline. Its founders believe that sociobiology occupies an intermediate position between the theories of Skinner, who assumes that man's behaviour is fully determined by his environment, and those of Konrad Lorenz, who argues that man

---

[1] Burrhus F. Skinner, *Reflections on Behaviorism and Society,* Prentice Hall, Englewood Cliffs, N. J., 1978. A critical analysis of "Skinnerism" is given in the works of: P. N. Fedoseyev, "Problemy sotsialnogo i biologicheskogo v filosofii i sotsiologii" (Problems of the Social and Biological in Philosophy and Sociology), in: *Voprosy filosofii,* No. 3, 1976; K. A. Shvartsman, *Novye tendentsii v razvitii sovremennoi burzhuaznoi etiki* (New Tendencies in the Development of Contemporary Bourgeois Ethics), Vysshaya shkola, Moscow, 1977; E. A. Arab-Ogly, *Otrecheniye ot sotsialnykh idealov. S chego nachinaetsya lichnost* (The Abandonment of Social Ideals. The Beginning of the Formation of the Personality), Politizdat, Moscow, 1979, and others.

[2] A detailed critique of modern biologism is presented in the work of Walter Hollitscher, an Austrian Marxist philosopher: Walter Hollitscher, *"Kain" oder Prometheus? Zur Kritik des zeitgenossischen Biologismus,* Verlag Marxistische Blätter, Frankfort on the Main, 1972; Walter Hollitscher, *Für und wider die Menschlichkeit,* Globus, Vienna, 1977. See also: Alexander Wernecke, *Biologismus und ideologischer Klassenkampf,* Dietz, Berlin, 1976.

is a prisoner of his aggressive instincts. R. Trivers, who is one of the leading theoreticians of sociobiology, even states that sooner or later political science, jurisprudence, economics, psychology and anthropology will all become sub-branches of sociobiology. Of course this has produced not only sympathy on the part of some scientists, but, and to a much greater extent, sharp criticism in which the excessive claims of sociobiologists are rejected, even though it is recognized that a number of their observations and hypotheses deserve close attention and an objective appraisal.

This is also true of the ideas of one of the founders of that approach—Edward Wilson, who presents them in *Sociobiology. The New Synthesis.*[1] In establishing a theoretical foundation for a sociobiology that would be the same for all forms of life, including man and human society, Wilson seeks to identify some common traits in the behaviour of all higher living organisms and to infer general governing principles. With specific regard to the nature of man, in another work, directly concerned with the subject,[2] Wilson develops the sociobiological approach in a way that stresses the determining role of biological (genetic) factors in higher manifestation of human behaviour. Lacking this, he believes, the social sciences would be quite limited and superficial descriptions of phenomena that do not involve their essence.

Wilson defines the philosophical and methodological foundation of sociobiology as a "new naturalism", and stresses that the latter seeks to prove its validity by generating two fundamental dilemmas. The first is that no living species, including man, possesses objectives extending beyond imperatives created by its genetic history. Wilson stresses that the human brain is constructed in such a way that this dilemma provides a fundamental constraint on its development and directs this development into a purely biological path. If, on the other hand, the development of the human brain is shown to be subordinated to the laws of evolution—to laws of natural selection—then even the selection of particular esthetic or religious convictions and

---

[1] Edward Wilson, *Sociobiology. The New Synthesis,* The Belknap Press of Harvard University Press, Cambridge, Mass., 1975.

[2] Edward Wilson, *On Human Nature,* Harvard University Press, Cambridge, Mass., 1978.

conceptions must be explained in terms of mechanical processes taking place in human brain. In other words, the human brain exists because it contributes to the survival of and increases in the number of genes guiding its activity. Thus human reason is a device designed to ensure the survival and reproduction of genes, rather than something else whose aim would be understanding the structure of atoms or self-cognition.[1]

As for the second dilemma, it is concerned, according to Wilson, with choosing among various ethical premises inherent in man's biological nature.[2] According to Wilson, there exist within the human brain certain "censors" that influence the ethical selection performed unconsciously by man; morality develops as an instinct, of sorts, under the influence of this selection process. In his opinion these "censors" are located in emotional centres of the human brain, most probably, under the cortex. Man's emotional reactions and moral conduct based on these reactions, according to Wilson, are to a certain extent preprogrammed by the course of the natural selections that has been taking place for thousands of generations. Accordingly, science faces the task of studying that genetic programme and the constraints that it imposes on the development of the human brain, that is, the task of recreating the evolutionary history of human reason. Such studies, in his view, constitute a logical complement of previous and current studies of man's cultural evolution. Their success would help us answer a question that constitutes the essence of the second dilemma, namely, which are the censors that we should obey and which of them should be constrained or else sublimated.

According to the sociobiological conception of genetic determinism, human culture cannot be viewed as some kind of superorganism developing on its own foundation. Rather, Wilson believes, cultural changes should be viewed as the median of the activities of a large number of persons seeking to adapt to social existence. From such a point of view, man's social evolution involves inheriting two types of attributes, namely, cultural and biological. And while cultural evolution, according to Wilson, can be viewed as Lamarckian and proceeds sufficiently

[1] *Ibid.*, p. 1.

[2] *Ibid.*, p. 4.

rapidly, biological evolution is Darwinian and proceeds more slowly. Thus, in his view, the Lamarckian proposition concerning the inheritance of acquired attributes may also be employed effectively in describing cultural evolution.[1]

Subsequently Wilson turns once more to the sociobiological conception of the origins of man's social behaviour, in which a key element is what he called the "auto-catalytic model" in his preceding work, *Sociobiology*. According to that conception, under the influence of natural selection genetic evolution increased man's capacity to create culture, while the latter, in turn, increased the genetic fitness of those who derived maximal benefit from it. In this connection Wilson considers that the influence of hereditary traits acquired by man from his ancestors on the course of cultural evolution is quite substantial. As evidence of this he points to the fact that whenever civilizations emerged their development followed a specific pattern as human societies grew in size and the complexity of their organizational structures increased.[2]

As we see, Wilson follows the approach of "genetic determinism", even though in a number of cases he does seek to distinguish the complexity of social structures in their evolutionary-genetic determination. "I interpret contemporary human social behavior to comprise hypertrophic outgrows of the simpler features of human nature joined together into an irregular mosaic... Others, such as religion and class structure, are such gross transmutations that only the combined resources of anthropology and history can hope to trace their cultural phylogeny."[3] Still, in his opinion even these "features" may become the object of statistical interpretation in the not too distant future.

He seeks to prove this by considering from the point of view of sociobiology what he believes are four basic categories of behaviour, namely, aggressiveness, sexuality, altruism, and religion. But the outcome of this examination points to the weakness and lack of prospects of the sociobiological approach, even though it is accompanied by interesting individual observations and inferences. It would appear that this is an unavoidable con-

---

[1] *Ibid.*, p. 79.
[2] *Ibid.*, p. 88.
[3] *Ibid.*, p. 95.

sequence of the basic methodological position, and of that relating to his world view, which make it impossible to move beyond what is already accepted at the beginning of the analysis as its premise and condition.

Evaluations of the scientific significance of sociobiology by both biologists and by philosophers, as well as by theologians and general opinion, are quite diverse. The continuing discussions in connection with sociobiology are often quite sharp and intense and involve decisive rejections of the whole theory or of its individual propositions as well as an acceptance of others. This may be seen, for example, in the debates in the American Association for the Advancement of Science that were reflected in the book entitled *Sociobiology: Beyond Nature/Nurture?*[1] Many authors, including Edward Wilson, Richard Dawkins and others, stubbornly defended sociobiology. A number of prominent scientists, however, also expressed highly convincing counter-arguments (e.g., Abraham Kaplan, Sylvester Gould).

The biologization of social phenomena acquires extreme forms when sociobiologists and other scientists who share their views make even wider generalizations and consider that the sphere of human culture generally somehow depends on genetic control. In a quite contradictory form this position is outlined in *Genes, Mind and Culture. The Coevolutionary Process* by Charles Lumsden and Edward Wilson.[2] In that book, which according to its authors fills a gap in sociobiology, an attempt is made to explain in terms of genetics the origin of the human reason and the diversity of cultures. In many cases, moreover, the authors seek to dissociate themselves from "traditional" sociobiology and stress that they are proceeding much further in their generalizations and conclusions. Seeking to create a theory of joint genetic-cultural coevolution, the authors develop a model that shows the correspondence of individual mental development with the development of culture, as well as the correspondence of the development of culture with genetic evolution.

---

[1] *Sociobiology: Beyond Nature/Nurture? Reports, Definitions and Debate.* Ed. by George W. Barlow and James Silverberg, Westview Press, Boulder, Colorado, 1980.

[2] Charles J. Lumsden, Edward O. Wilson, *Genes, Mind and Culture. The Coevolutionary Process,* Harvard University Press, Cambridge, Mass., London, 1981.

The point of view according to which the cultural behaviour of human beings is seen as a product of their biological development makes it possible, in the authors' opinion, to develop a fresh perspective on the social sciences. In their view it may prove to be true that even though *homo sapiens* is indeed the most complex species living on Earth, it is still not as complex as modern social theories surmise. In rejecting the biological, natural-scientific approach to the analysis of man in terms of neurophysiology and experimental psychology, Lumsden and Wilson also adopt a critical view of structuralists, who generally study complex phenomena that are specific to man alone—such as the qualitative characteristics of language, the stages of intellectual development which make up a "genetic epistemology" (Piaget), the origins of myths and rituals (Levi-Strauss and others). Finally, the authors believe that proponents of psychoanalysis address themselves to even less tangible aspects of the activities of the brain. Lumsden and Wilson believe that structuralism and psychoanalysis are not science in the sense attached to that term in natural science, but rather mere descriptions.

As for Marxism, it is their view that "it is more like Lamarckism and ontogenesis, competing theories that correctly described certain outward features of evolutionary change but postulated erroneous underlying mechanisms". This "error" is ascribed to its "tendency to conceive of human nature as relatively unstructured and largely or wholly the product of external socio-economic forces".[1] Lumsden and Wilson also seek to present this idea in their new book entitled *Promethean Fire. Reflections on the Origin of Mind*,[2] in which they carry this idea to the absurd assertion that Marxism views the mind as "biologically unstructured"[3]. Of course, as we have already seen, such assertions have nothing in common with the actual facts.

Lumsden and Wilson did not take the trouble to analyze more

---

[1] Charles J. Lumsden, Edward O. Wilson, *op. cit.*, p. 355.

[2] Charles J. Lumsden, Edward O. Wilson, *Promethean Fire. Reflections on the Origin of Mind*, Harvard University Press, Cambridge, Mass., London, England, 1983.

[3] *Ibid.*, p. 169.

closely the scientific method of Marxism and the solution that it provides to the problem of man, including the question of the relative role of social and biological factors in man's formation and development. They make use of the distorted conceptions of Marxism that are current in bourgeois literature.

Thus, the claims of "non-traditional" sociobiology of "theory of genetic-cultural coevolution" as presented by Lumsden and Wilson are quite far-reaching and in many cases are directed against the Marxist teaching concerning society and man. But what is needed, I believe, is not only a general methodological critique of the "theory of genetic-cultural coevolution" but also a close analysis of the facts that are adduced in its substantiation. Already in the first reviews of Lumsden and Wilson's book it was noted, for example, that the authors' conclusions "depend on arbitrary assumptions and mathematical analyses that are ambiguous. The validity and generality of their assumptions about the inheritance of social behaviour must be tested empirically... Most critically, their theory can have little heuristic value because it is open to only weak tests at best."[1]

Such a conclusion is quite symptomatic as a recognition of the appropriateness of a careful verification of the factual basis of the "theory of genetic-cultural coevolution". At the same time it is important, in my view, to develop on the basis of Marxist methodology wide-ranging concrete and positive studies of the interdependence of social and evolutionary genetic factors in the formation of man's individual and historical development, including manifestations of his higher spiritual properties recorded in culture.

Marxism recognizes that at various stages of man's development, biological factors, operating in a mediated and transformed form, have varying degrees of importance. It is not there, however, that man's essential forces originate, and it is not on the basis of such an approach that one can identify the concrete mechanisms, or the conditions of and means for their operation in man's individual and historical emergence and development. These mechanisms and conditions are already outlined in Marx's relevant proposition that society is man himself in his

[1] C. Robert Cloninger, Shozo Yokoyama, "The Channeling of Social Behavior", in: *Science,* Vol. 213, No. 4509, 14 August 1981, p. 751.

social relations.[1] This emphasizes a dialectical interdependence between the formation and development of man as an individual and a personality and his history as the progress of human society, since in both instances the same social governing principles operate.

According to Marx a central role in that interaction is played by the object-oriented activity of man as a social being. Man's individuality and uniqueness are manifested in the organic unity and integrity of the development of the totality of his needs and abilities, which are formed through object-related forms, modes and means of culture—labour, language, skills, aptitudes, knowledge, artistic creativity, etc. According to Marx, "man appropriates his comprehensive essence in a comprehensive manner, that is to say, as a whole man."[2] All of his human relations to the world—vision, hearing, smell, touch, thinking, contemplation, feeling, desire, activity, love—are, he emphasized, "in their *objective* orientation, or in their *orientation to the object,* the appropriation of the object, the appropriation of *human* reality".[3]

This thought may be viewed as central in understanding man's individual and historical development. It also solves the problem concerning the correlation therein of social and biological factors. The entire process appears as the appropriation and reproduction, respectively, of mankind's socio-historical experience, of his material and spiritual culture.

Characteristically, precisely this approach in solving these questions was employed by Antonio Gramsci, a prominent Italian Marxist, in criticizing the excesses of both anthropologism and biologism in understanding man. Gramsci also opposed vulgar sociologism which seeks to explain any human behaviour in terms of the influence of external conditions of the social environment, defended Marx's teaching concerning man's essence as the totality of all social relations, and saw the way to man's emancipation and development in a change in these re-

---

[1] See: Karl Marx, "Economic and Philosophic Manuscripts of 1844", *op. cit.,* p. 299; Karl Marx, "Theses on Feuerbach", *op. cit.,* p. 4.

[2] Karl Marx, "Economic and Philosophic Manuscripts of 1844", *op. cit.,* p. 299.

[3] *Ibid.,* p. 300.

lations and in active involvement of the personality. Gramsci saw man as a series of active relations and a process.

In Soviet literature such a historical approach which explains the social mechanisms, conditions and means of man's formation, was developed on the basis of the works of Karl Marx in the 1920s and 1930s first by Lev Vygotsky,[1] then by Sergei Rubinshtein,[2] and subsequently by their followers, particularly by Alexei Leontiev.[3] On the basis of Marx's theory concerning the social essence of man and the forming influence on the latter of object-oriented activity, in the course of which the "appropriation of human reality" takes place, the Soviet scholars have shown that the historical approach plays a leading role in understanding the mechanisms, conditions and means of the formation of the personality. That approach is in opposition to biologistic and naturalistic conceptions, and places into the foreground the idea of a social, cultural-historical development of man, including the transformation of his biological nature as a result of that development.

Such a historical approach clashes not only with conceptually unfounded attempts to oppose to it the "teaching" concerning "the two programmes" in human development, namely, the genetic and "non-genetic" (social, cultural, etc.) programmes, but also with attempts to sharply dissociate from it such scientific advances as, for example, Ivan Pavlov's theory concerning higher nervous activity, modern perceptions of the brain's systemic activities, biological studies of reasoning activities, evolutionary-genetic prerequisites of human behaviour, etc. Such attempts, occasionally carried to the level of "pure sociologism", are hardly productive, and many researchers, including those developing the legacy of Vygotsky, do not accept them. For example, Leontiev emphasizes the absence of grounds for dividing psychological problems into two types, namely, problems

---

[1] See: L. S. Vygotsky, *Izbrannye psikhologicheskiye issledovaniya* (Selected Psychological Studies), Nauka, Moscow, 1956.

[2] See: S. L. Rubinshtein, *Osnovy obshchei psikhologii* (Principles of General Psychology), Uchpedgiz, Moscow, 1940.

[3] See: A. N. Leontiev, *Deyatelnost. Soznaniye, Lichnost* (Activitv Consciousness. Personality), Politizdat, Moscow, 1975; *idem., Problemy razvitiya psikhiki* (Problems of Psychological Development), State University. Publishers, Moscow, 1981.

that are to be elaborated on the basis of Pavlov's theory, and those that are to be approached on the basis of historical materialism. "The actual problem," Leontiev notes, "is, of course, the converse one, namely, to apply a *common approach* to all problems of human psychology, and in this way to include them into the system of a unified science."[1]

The task formulated by the prominent Soviet psychologist, namely, to develop a common approach and a unified science system may well be of still wider significance, for it extends to the Marxist theory concerning man, including the problem pertaining to the relative role of social and biological factors in man's existence and development.

Since "the individual *is the social being*", it follows that "his manifestations of life—even if they may not appear in the direct form of *communal* manifestations of life carried out in association with others—*are* therefore an expression and confirmation of *social life*".[2] Through the intermediary of social communication, human activity, in its mediated—either object-oriented or linguistic—forms, creates the condition and means of man's individual and historical development as a social being. Accordingly, in the course of man's education and training the appropriation of the human race's socio-cultural experience operates as a source of man's development. It follows that individual psychological activity is "predefined" in social communication and collective activities. In the works of many Soviet scholars—psychologists, educators and philosophers—that approach is substantiated with fundamental research (cited, in particular, in the previously mentioned works by Alexei Leontiev as well as those by Vasily Davydov, Vladimir Zinchenko, and still others).

Obviously, in this context Marxists do not base their analysis on some kind of a standard as regards man and personality that would presuppose an absolute equality of abilities and manifestations of individuality. As is known, Lenin emphatically rejected such an equality in abilities or similarity in physical strength and mental ability. He stressed that "in this respect men are *not*

---

[1] A. N. Leontiev, *Problemy razvitiya psikhiki,* p. 363, (my italics—*I. F.*).

[2] Karl Marx, "Economic and Philosophic Manuscripts of 1844", *op. cit.,* p. 299.

equal ... when socialists speak of equality, they always mean *social* equality, equality of social status, and not by any means the physical and mental equality of individuals".[1] The latter exists in the form of certain "potentialities", but their realization and development also require specific *social* conditions.

The historical approach (though obviously not its narrow sociologistic interpretation) makes it possible to perceive the dialectics through which the social mediates and transforms the biological and see the "humanized nature" and "human sensibility" as the outcome of all previous history.[2] It seems to me that such an approach also makes it possible to arrive at the correct view of man's future prospects, including natural-biological ones, which must therefore be understood as humanized, that is, mediated and transformed as a result of human activity and of the human race's socio-cultural progress.

It is on such a basis, when man becomes, in the words of Marx, an "end in itself" of social development that is turned towards a communist perspective, that Marxists define the task of forming the new man, one of the central tasks under socialism. Its accomplishment must be inseparably linked with further advancement of society along the path of socio-economic and scientific and technological progress, and with the development of all vital spheres of the social organism. Characteristically, not only in practical activities, but also in science, including the study of Nature, seemingly remote from this range of problems, the inner logic of development and the influence on it of social needs and tasks lead it to man, and the cognition of man acquires a role which is central to the entire system of scientific cognition. Moreover, science changes substantially in a social and human sense, in the foundations of its world view and value orientations. These processes are so fundamental and long-term in nature that they largely determine the main directions of scientific development and form a new type of social and human functioning of science.

---

[1] V. I. Lenin, "A Liberal Professor on Equality", *Collected Works*, Vol. 20, Progress Publishers, Moscow, 1972, pp. 144, 146.

[2] Karl Marx, "Economic and Philosophic Manuscripts of 1844", *op. cit.*, pp. 301, 302.

CHAPTER ONE

SCIENCE AT THE SERVICE OF MAN AND MAN AS THE "MEASURE OF ALL SCIENCES"; SCIENTIFIC COGNITION AND HUMANISTIC VALUES; THE ETHICS OF SCIENCE

> Science must be a servant to man.
>
> *Nikolai Chernyshevsky*

> What does the evolution of the human race in its most recent past teach us, in what direction is it proceeding and what forces is it placing in the foreground?—SCIENCE AND DEMOCRACY.
>
> *Kliment Timiryazev*

The words cited above are those of Russian thinkers—a philosopher and a scientist—who already towards the end of the last century and the beginning of the present one were able to clearly express not only the role to be played by science subsequently, but also its close association with man, with social progress, and with advanced forms of social existence. Today that interdependence has increased as never before. Many aspects of the present existence of the human race would literally be impossible without scientific discoveries, without science's intrusion into the material and spiritual foundations of human civilization, while one of the factors of intensive scientific progress in the twentieth century has been the influence on science of new elements of social life resulting from the development of large-scale industries and from socio-cultural progress.

Today the unlimited possibilities of scientific and technological progress have become fully evident, as has its social and humanistic meaning. But certain other phenomena that are dangerous for the future of the entire human race have also become more discernible. These are associated with specific characteristics of the early phases of the scientific and technological revolution and its development in the context of the struggle between the two world social systems. Numerous global problems have either emerged or else become more intense. In that connection, as never before, the organic interdependence of scientific and technological progress with the social conditions in which it takes place, as well as with man himself as the centre of intersection of the basic "lines of force" governing the development of science, technology, culture and overall world history, has become apparent. It has become evident that man's future will be largely determined by the totality of these factors, within which science plays a most important role.

All this has greatly enhanced interest in science and technology, which, together with other basic factors of social development, are now being subjected to humanity's moral judgement. A humanistic, moral evaluation of the role and significance of scientific progress in the life of individuals and of the human race is one of the symptoms of sharp and intense concern with which we are approaching the threshold of the third millennium.

Is science a force that serves the needs and interests of human beings regardless of the type of society in which it functions and develops, or does society fully determine its direction, including possible negative manifestations? Can science contribute, and if so, to what extent, to man's development or does it threaten it? How does it relate not only to the social conditions and factors influencing its development but also to values, and to humanistic values in particular, and what role can these play in its development? Finally, how does science influence changes in the human race's humanistic principles and to what extent does it depend on such changes itself? How does man change in the course of scientific and technological progress and how is that progress transformed under the influence of changes in man, in his civilization, and his culture?

Today these questions hold a central position in philosophic discussions; yet the latter, in turn, also depend on certain needs defined by life itself and by the development of social relations. Naturally, these questions are approached in different ways in different social systems, even though there are many common, global traits in the way in which they are formulated. In the theoretical solution of these questions, a role may also be played by philosophy which examines their aspects that relate to world views and analyzes humanistic problems of the present and the future. Today this examination and analysis perform not only a constructive scientific and social function, but also a function of criticism, as they have a role to play in the ideological and political struggle taking place in the world.

1. Science in the Present World and the Present World of Science: Its Transformation into a Direct Productive and Social Force; the Emergence and Intensification of Global Threats and Problems; the Dilemma of Scientism and Anthropologism; Alarmism, Utopian Optimism and Scientific Humane Realism

As we turn to an analysis of modern science we face at least four interrelated questions:

(1) What is science and how does it originate and function as a specific activity guiding man in the world?

(2) Why does science emerge and develop in particular forms rather than others at various stages in history?

(3) For what purpose does science exist, in view of its relation to man and society?

(4) (This question has emerged quite recently). What are our possibilities in regulating science and in what way can we make use of them?

Answers to these questions underlie various aspects of the study of science: socio-philosophical, historical, sociological, value-related and axiological (humanistic), and, finally, organizational and administrative. And while the basic emphasis will be given to socio-philosophical, as well as to value-related and axiological-humanistic problems, I will seek to examine them in their relationship and interaction with other forms of study of science.

Today scientific knowledge has become a necessary component in the functioning of human society, many aspects of which were brought to life by scientific discoveries. Modern science has radically restructured traditional spheres of production, while some industries were born in laboratories (the atomic industry, radio electronics, space technologies, the creation of polymers, biotechnologies, etc.).

The revolution in science and technology has enabled the human race to take a vast leap forward in a relatively short period of time in mastering Nature's secrets and turning them to practical use: the theory of relativity, quantum mechanics, nuclear physics, the chemistry of polymers, cybernetics, molecular biology—all these scientific disciplines, which define the face of modern science, have emerged and developed in this century. Moreover, their most significant achievements were made in recent decades. According to available estimates, the volume of scientific information and discoveries produced during that period has exceeded all that science had achieved over the thousands of years of its existence, and is continually increasing.

Scientific and technological progress is gaining momentum. In particular, the delays between the scientific discovery of some principle or law of Nature and its practical utilization are steadily decreasing. As these delays are reduced, science increasingly precedes material production within the latter's sphere itself. It transforms and organizes production on the basis of principles and laws of Nature that are first established theoretically, rather than only affects from the outside the forms of production that have developed in unorganized ways and on the basis of empirical experience. The economic effectiveness of capital investments into scientific research serving production activities, moreover, has been found to exceed that of investments in other components of the latter. There is no need in the present context to cite specific figures, for all this is widely known. Generally it may be said that today the transformation of production activities through the utilization of science has become a new powerful motive force in the production process.

It is important to note that social sciences, too, are becoming involved in that process. This is an expression of an effective tendency whereby natural sciences are combined and interwoven with social sciences, as dictated by the development of the social nature of labour. As the sphere of social applications of the findings of natural sciences to production activities widens (in such projects, for example, as those in nuclear energy, the transformation of deserts, space research, complex automation, etc.), a growing need arises for specific forms of social knowledge concerning the conditions and consequences of such applications, while the dependence of natural science itself as a productive force on the availability and comprehensiveness of that knowledge also increases. Today teams of economists, sociologists and psychologists are being assigned to scientific projects and to their practical implementation, and the character of their work and criteria of scientific validity of its results must increasingly approach the precision of natural-scientific knowledge.

Lenin wrote: "It is common knowledge that a powerful current flowed from natural to social science not only in Petty's time, but in Marx's time as well. And this current remains just as powerful, if not more so, in the twentieth century too."[1] Integration processes possess objective foundations that lie within modern science itself and at the same time are dictated by practical requirements that are associated with profound and long-term processes of social change taking place in connection with the development of science.

The search for scientific knowledge is now no longer limited to one or several focal points in some particular science. It is more like an outline tracing the major directions of scientific and technological progress in a space occupied by numerous sciences, but most often at "interfaces" among them. This is why today scientists are increasingly aware of the need not only for integrated approaches to complex problems, but also for creating wide-ranging and ramified research programmes that presuppose a concentration of substantial material and human resources, a deep linkage of science with practical activities and with man himself, and harmonization of science with all of modern culture.

The integration of sciences is determined not only by their own internal requirements, but also by requirements associated with the transformation of science into a direct productive force in society. In analyzing the socio-economic substance and consequences of that phenomenon Marxism has shown that the latter is by no means the result of inherent development of the content of scientific knowledge, but rather only an aspect and manifestation of the process of further socialization of production and of human activities generally; moreover, the cooperation and social character of labour are becoming a powerful productive force in themselves.

---

[1] V. I. Lenin, "Socialism Demolished Again", *Collected Works,* Vol. 20, p. 196.

The new role of science is an expression of the social potentialities of modern production activities, as reflected above all in the transformation of the structure of labour, and of the sphere of material production generally. Marx noted that with the socialization of production labour is increasingly influenced by the material force of knowledge. "The implements of labour, in the form of machinery, necessitate the substitution of natural forces for human force, and the conscious application of science, instead of rule of thumb."[1]

Scientific work also experiences profound changes. As the revolution in science and technology proceeds, research changes its social form and is drawn closer to the labour of workers in the productive sphere. While remaining an intellectual activity aimed specifically at obtaining new knowledge, scientific work becomes productive as regards its form and conditions. The latter thus become part of the production process and are "embedded" in it, as it were. Researchers now use large-scale industrial research equipment and work collectively. Large numbers of scientists work at one location; there is a division of functions among them, and the results of their work are embodied in material products. As a result of the penetration of production forms into engineering experimental infrastructures and into the organization of science, which lends it traits of a socialized production process, the following changes have taken place in scientific work.

First, the labour of scientists and science itself are becoming industrialized (due to changes in the nature and forms of instruments of the experimental influence of modern science on Nature). The findings that constitute the basic content of scientific discoveries in the twentieth century, and above all those in basic sciences, were all obtained due to the development and use in research of the most complex, expensive and sometimes huge machines and devices: nuclear reactors, accelerators of atomic particles, space probes, satellites, computers, aerodynamic and hydrodynamic facilities, the most sensitive recording devices, space communications systems, etc.

Secondly, the role of individual scientific labour has changed as scientific activities have acquired a collective and mass char-

[1] Karl Marx, *Capital,* Vol. 1, p. 364.

acter. Modern large-scale industrial-type facilities designed for carrying out scientific experiments and research bring together large collectives of scientists, engineers, technicians, laboratory assistants, skilled workers, and require cooperation and complex division of labour among them.

Thirdly, the character of many modern scientific tasks and projects has become such that only major social funds or the effort of the entire society, consciously mobilized in a planned manner, can be effective.

Still, all the deep underlying processes associated with the transformation of science into a direct productive force in modern societies exist and are manifested within specific systems of socio-economic relations. In actual fact, science operates as a social institution that depends on these relations. Accordingly, in order to determine the conditions, forms and consequences of the functioning of science in a society it would certainly be insufficient to merely examine the internal processes of modern industrial production and the corresponding potentialities of science. The preceding reflects the essence of the Marxist approach to such a complex and contradictory phenomenon as modern science. And it is in this, it would appear, that there is a key to a correct understanding of the mechanisms governing science's social functioning and its influence on the life and the future of man.

In summarizing, it should be added that in Marxist-Leninist theory there exist a well established tradition and certain priorities in studying the development of science and technology. This should be especially emphasized, as it is often asserted that Marxists, and particularly Soviet theoreticians, have only turned to the problems of the scientific and technological revolution in recent decades. It is true that it is in the latter period that a large number of studies have been carried out that examine the processes of the development of modern scientific and technological progress from various angles. It is well known, however, that already in the last century Marx established the basis for studies of this type and formulated the basic principles of what can be presented today as the *Marxist-Leninist theory of the scientific and technological revolution*. For it was Marx who established that science is not an isolated phenomenon but is a specific social institution in modern society.

This discovery makes the essence of modern scientific and technological progress more understandable.

Further, Marx also identified the dominant tendency in the development of science as it is being transformed into a direct productive force in society. That scientific prognosis, made as early as in the middle of the last century, is being fully confirmed today before our eyes, and, in effect, characterizes one of the specific features of modern scientific and technological progress.

Finally, Marx demonstrated that the development of science and technology as they interact with the development of productive forces is a major element of a social transformation of the type which is a necessary condition for man's free and comprehensive creative development. Thus outlined are the future prospects of scientific and technological progress, within which man is the main factor.

Marx noted that at a certain stage in the development of productive forces in society science becomes a *"direct productive force"*[1] while man's own universal development becomes a condition for a universal development of productive forces.[2] The linkage between the development of productive forces, progress in science and technology, on the one hand, and man's development, on the other, brings out perhaps best of all the specific features and the qualitative characteristics of the Marxist-Leninist conception of the scientific and technological revolution. At the same time, it is a stable point of reference for the development of both society as a whole and science and technology in particular. That point of reference is provided by the orientation of society and science towards free and comprehensive development of man.

Such an orientation makes it possible to understand much that has taken place in modern society and to correctly evaluate not only positive but also negative facts, and thus not only to see the benefits of the current stage in the development of science and technology but also to assess the negative phenom-

---

[1] Karl Marx, *Grundrisse der Kritik der politischen Ökonomie (Rohentwurf, 1857-1858)*, Verlag für fremdsprachige Literatur, Moskau, 1939, p. 177.

[2] See: Karl Marx, *A Contribution to the Critique of Political Economy*, p. 202.

ena that are associated with it. Most importantly, such an orientation reveals the actual potential and the long-term outlook for overcoming these negative phenomena as society develops in the direction of a new and more perfect condition, socialism and communism.

To the greatest extent the above applies to the complex of global problems whose emergence and intensification has largely been caused by the intensive development of science and technology and by the increasing industrial utilization of their findings on the basis of traditional technologies. The current revolution in science and technology has brought about unprecedented increases in labour productivity and the scale of production, as well as unexampled results in mastering the forces of Nature: the secrets of the atom and the mysteries of outer space, the depths of the world's oceans and the microcosm of living cells have become more accessible to the human mind equipped with science. Yet these advances were also accompanied by setbacks as the danger of military applications of technological findings of science has increased, as has that of depletion of natural resources, the energy and food crises, etc. Serious distortions have also taken place in modern culture and education as the gap between the "two cultures" (C.P. Snow)—scientific, technological and humanistic—has become deeper. There emerged a mortal threat to the biological existence of the human race itself, for not only modern nuclear missiles but also their production, which intensifies the pollution of the environment by substances dangerous to human health, can produce irreversible consequences that would be fatal for the human race.

With the intensification of global problems, the future of the human race is increasingly linked to the possibility of their resolution. Scientific and social forecasts of the future are now turning to global problems in a search for real alternatives as regards human civilization and future prospects for man. Unfortunately, this problem area also gives rise to a mythology that capitalizes on the prestige of science while having nothing in common with it. Global problems are considered in isolation from the basic contradictions of the era of transition from capitalism to socialism and represented as something "suprasocial" and "above social classes"; it is claimed that their solution is possible only within the scientific and technological sphere (converted into a one-sided absolute), or else by changing man himself. The latter is then recognized as the major prerequisite and condition for solving all other problems.[1]

Today's global problems exist within an integral system; they are dialectically interdependent, which is a manifestation of their social nature. It is precisely this factor that makes them necessary components of all projections of human civilization and of man himself. This is why we refer to: (1) *inter-social* global problems, relating to mutual interactions among such social communities as socio-economic systems, states, etc. (These problems include peace and disarmament, world social and economic development, overcoming the backwardness of various countries and regions, etc.); (2) *anthroposocial* global problems associated with relations between man and society (scientific and technological progress, education, culture, population growth, health, biosocial adaptation of man, and his future) (3) *natural-social* global problems pertaining to the interaction of man and society with Nature (natural resources, energy, food, the environment, etc.).

All these problems influence the future of human civilization in varying degrees. Moreover, their influence is often most direct, and permitting for no delays in reducing corresponding dangers. Naturally this applies above all to problems of peace and disarmament, which, unfortunately, remain the primary concern of the human race, and may even be described as the "tragic syndrome" of human civilization throughout its development and especially today, when its very existence is threatened. The latter situation is due both to the fact that, as a result of the development of the negative consequences of the internationalization of social life, wars are also acquiring a dangerous tendency to become global, as well as to the unprecedented increase in the destructive potential of the means of warfare, the use of thermonuclear weapons, missiles, etc.[2] A new

---

[1] Further details in: I. Frolov, *Global Problems and the Future of Mankind,* Progress Publishers, Moscow, 1982.

[2] The Swiss researcher Jean-Jacques Babel has estimated that during the preceding 5,500 years in the history of human civilization there have been 14,513 wars, in which approximately 3.5 billion persons lost their lives. During the Second World War, when the hostilities involved the

thermonuclear world war can exterminate all life on Earth and transform this blue planet into a lifeless celestial body contaminated by radioactive waste. Consequently, in the age of nuclear missiles a world war is absolutely inadmissible.

If a peaceful, secure future for all mankind were to be guaranteed, humanity would be free from the heavy burden of the arms race imposed by imperialism. Disarmament is an urgent imperative of the modern age. The solution of this major global problem would ultimately reduce the intensity of the entire range of problems plaguing the world, including economic, social and ecological ones.[1]

Today global problems affect different countries and regions of the world unequally. In the long term, however, the intensity of individual problems at the global level and the national priorities as regards their solution may change quite substantially already by the year 2000. This is revealed, for example, by consideration of the problem of natural resources, upon the solution of which the economic growth, social development, and more generally the future of the human race greatly depend. That problem is even more acute if one takes into account the fact that the natural energy resources are close to being depleted today. The problem is to gain access to new natural resources, including energy resources, and utilize them most effectively. The latter requirement, however, is also associated both with the level of development of science and technology and with social factors, as well as with relations between states that own and that consume natural resources. Consequently, the problem of natural resources is not a purely scientific or technological problem, as some bourgeois ideologists claim: in the final analysis, it is determined by a number of socio-economic and political factors; therefore, the impending "global crisis"

---

territories of 40 states, a total of 50 to 55 million persons perished while the number of wounded was 34 to 35 million persons (of the latter, 20 to 25 million remained disabled).

[1] According to United Nations data, total military expenditures in the world constitute more than $650 billion per year, several times as much as the expenditures for health and education. The rechanneling of a major share of world military expenditures to other sectors, according to the UN experts, could increase rates of economic growth and bring an end to famine, disease and illiteracy.

in that field, concerning which there is so much discussion, possesses purely social origins.

Economic growth, in many respects influenced by the growing energy-intensity of labour, has led to a situation in which the production of all types of primary energy has vastly increased.[1] This makes the energy problem one of the most acute of today's global problems. Capitalist countries, moreover, seek to overcome the energy crisis not so much by relying on internal resources as by continuing their neocolonialist exploitation of the developing countries.

Obviously, in the socialist societies the energy problem has also become more pressing in recent years, leading to increased prices for energy resources. But in their case the problem is approached on a fundamentally different basis and being solved, in particular, within the framework of the Council for Mutual Economic Assistance which provides for equitable and mutually advantageous cooperation of these countries within a joint energy programme. In its context, accelerated development of nuclear power generation has been defined as one of the major approaches to solving the fuel problem.

The problems of nuclear power have acquired a major role in the energy programmes of many countries and are now truly global in nature.[2] Obviously, nuclear power would not provide either today or in the near future (at least not before the third millennium) an alternative to "traditional" energy resources, and as many research scientists have rightly stressed, the human race is not yet condemned to choose between "the atom and the candle". Large-scale and more effective utilization of all types of natural energy sources (including solar energy) will contribute to man's reaching new frontiers of social and economic progress, and that, in turn, will exert a feedback influence on the strengthening and development of the energy basis of man and Nature.

---

[1] Available estimates indicate that it grew from 2.7 billion tons of reference fuel in 1950 to 9.0 billion tons in 1975, i.e., by 3.3 times. By the year 2000 one may expect a substantial growth in the absolute level of consumption of energy resources to approximately 20 billion tons of reference fuel.

[2] It is expected that the share of nuclear fuel within the global fuel balance of electric power stations may reach 45 per cent by the year 2000 and 60 per cent by the year 2020.

Still, a large-scale industrial application of nuclear power, while reducing the pollution of the environment by combustion products (carbon dioxide, sulphur, etc.) produces a different danger, namely, environmental pollution by radioactive by-products of uranium and plutonium fission. Much is being done throughout the world to prevent this danger and rule it out in the future.

The situation is also urgent with regard to the food problem, that is, providing the energy resources that the vital activities of the human organism require.[1] Bourgeois theoreticians seek to explain the intensification of the food problem in the world in terms of "natural" causes—such as a population growth that overtakes increases in the production of food, a shortage of arable land suitable for cultivation, declining soil fertility—and also in terms of the technological backwardness of developing countries. Differences in levels of income and levels of food consumption between various social layers in the developed and developing countries are not taken into account. In actual fact, the shortage of food in the world is relative, rather than absolute, and is attributable not to "natural" or technological factors, but above all to social and economic factors that impede, among other things, the introduction of advanced cultivation methods which could resolve the food problem on a world scale.

Suffice it to note that today's agricultural output is only 3 to 4 per cent of what could be produced with effective application of modern science and technology in all countries. Their use would vastly improve crop yields and animal husbandry productivity. Still ahead of us are new prospects being opened by the development of science and technology for peaceful uses: the efforts of researchers are bringing closer the age of abiotic food production, without the use of plants or animals. In terms of their significance for the destiny of human civilization, fu-

---

[1] Each year as many as 10,000 persons in the world die of hunger. On the average, 16 per cent of the population of the globe suffer from malnutrition (3 per cent in developed countries and 25 per cent in developing states). It will be necessary to at least double the output of grain crops in order to feed the Earth's population by the year 2000. If one also considers the need to overcome shortages of calories, world food production must triple during the next 20 to 25 years.

ture advances in world production of artificial foods are sometimes compared to the introduction 6,000 years ago of agriculture, which led to the emergence of the first highly developed cultures—Sumer, Babylonia, and Egypt.

The environmental problem, with its close causal links to problems of economic growth, scientific and technological progress, natural resources, energy, and food, is also profoundly social in nature. If one keeps in mind the way in which problems of the environment are reflected in mass consciousness, their role today is perhaps even more important than that of many others, and understandably so, for facts concerning the destruction of the natural environment are obvious and accessible to all. But it is more difficult to understand the profoundly social essence of this problem and to see that it can be resolved only as a consequence of fundamental social, cultural, economic, technological and axiological shifts that are needed to overcome dangerous disbalances in man's reciprocal relations with Nature.

In overcoming the environmental crisis the human race will have to solve many other complex global problems. They include those that result from population growth and other current and future demographic processes. Of course, the situation in this regard is quite complex, and many scientists advocate active and effective demographic policies related to general processes of socio-economic development and to the specific social organization of individual countries and their dominant ideologies and traditions.[1]

As we turn to that global problem we would perhaps perceive most clearly that the common destiny of the human race

---

[1] While the rate of its growth has recently begun to decline, world population is expected to increase by 2.26 billion persons by the year 2000, i.e., by 50 per cent, to 6.35 billion. In short, during the last 25 years of the 20th century it will have increased by as much as during the first 1,950 years of the modern age. Nearly 90 per cent of the additional population, moreover, will live in the developing countries, where the rate of growth will be 4 times as high as in the developed states. As a result, assuming that substantial differences in standards of living will continue to exist, by the year 2000, 78 per cent of the world's population will live in inadequately provided regions as against 68 per cent in 1950.

is inseparable from that of each individual. In this connection, additional questions arise: what will be the future of man; what are his prospects in the context of the processes taking place today in the social, economic, natural-biological and cultural spheres of human civilization? How does all this interact with the complex of global problems and with the ways in which these may be resolved in the future?

The first point that should be noted with regard to the problem of man and his future concerns the fundamentally new conditions of man's existence and development that have largely been created by scientific and technological progress and the emergence of global problems. These new conditions express themselves, first of all, in changes in the labour activities of individuals. Secondly, they are characterized by changes in man's direct relations with the natural environment. Third, they are associated with new factors in the vital activities of man as a biological species. Finally, these new conditions are expressed in changes in man's interrelationships as a personality with both society and other individuals whose aggregate constitutes the human race.

The number of factors exerting a negative influence on man's living and working environment has already reached dangerous thresholds that threaten man's future. Above all these include various chemical factors (the human race has now available to it more than 5 million chemical compounds that are used in a wide diversity of spheres); increases in background radiation that have led to increased incidence of disease (approximately 20 per cent of all diseases today are attributable to pollution of the environment); and intensification of mutation processes resulting in the birth of larger numbers of children with hereditary defects and a greater risk of oncological diseases. At the same time, urbanization produces an intensification in the rate and rhythm of everyday life and in nervous and psychological tension in the course of labour activities and interpersonal communications. These produce stress and an entire range of "illnesses of civilization", including cardiovascular diseases and psychological breakdowns.

Thus the fundamental questions are: will man be able, in a historical perspective, to survive as the *homo sapiens* species? Is he not threatened, as many scientists assert today, with biological or genetic degeneracy? Perhaps certain measures to prevent this should already be adopted today? Finally, is it possible to intrude into sensitive processes of man's biology and genetics without damaging human nature, what are the limits of acceptable intrusion, and is it actually necessary in the first place? Answers to these questions are already relevant today, and they will play a major role by the year 2000.

Science seeks not only to facilitate better adaptation of man to the new conditions caused by the intensification of global problems, but also to develop new methods for overcoming, or at least reducing to the minimum, the negative consequences of factors that threaten human existence. It discovers new ways of acquiring knowledge and of carrying out practical activities, ways associated with the emergence of genetic engineering and biotechnologies, human genetics and medicine. It studies the potential of the brain, develops microelectronics, seeks to create an "artificial intellect" and to contribute to man's psychophysiological development and increasing his longevity.

This also raises, however, an entire range of complex socio-ethical and humanistic problems relating to controls over this type of research both today and in the future, since it involves dangerous experiments on man that may infringe upon his dignity and freedom. More clear-cut socio-ethical and humanitarian regulation of research is needed on a global scale, as is the development of ethical codes and international legislation that would recognize the need both for further progress in scientific research relating to man and for possible rigorous constraints, so as to preclude manipulative inhumane approaches.

Scientific cognition of man, however, is not limited to the study of his biology, genetics, and psychophysiology. It also has a broader context which includes behavioural sciences and the system of education and training that form individuals as unique personalities. The cognition of man as an individual and a personality requires an integrated approach based on the principles of new, real humanism capable of uniting all of progressive mankind in an active effort at the individual and social levels for the benefit of man, his wellbeing and his future.

The problem of man and of his future thus acquires distinct significance among other global problems. More, it is the central point, as it were, of the entire system of these problems. The general approach and strategy for solving global problems will largely depend on the way in which the problem of man and his future will be solved.

Such an approach to the subordination of these problems reflects Marxism's humanistic world view and social position. This may be called the scientific philosophy of global problems. Today one already clearly discerns their influence on virtually all aspects of human life and that of the human race, such as the spheres of material production, culture, politics and ideology, world views and ethics.

Advances in the revolution in science and technology and the appearance and intensification of global problems have led to the emergence of a new social and spiritual environment within which many questions are being defined and resolved today, including some that bear directly on the human race's future. Unfortunately, today the matter at hand is often not only that of improving the conditions of human existence but, above all, of guaranteeing the survival of the human race during the next several decades. Is such a formulation of the problem justified? Is there not an overdramatization of the actual situation in the world?

Marxists have provided a scientific answer to these questions that, on the one hand, takes into account the novel character of the present situation and the importance of solving the whole range of modern global problems in the context of general social transformations in the direction of socialism and communism, and, on the other, places even greater emphasis on the need for these transformations. The basic principles of a scientific understanding of the human race's future—which also underlie real progressive development of a substantial part of the human race living in socialist countries—make it possible to take a critical assessment of a variety of projects that are not related to such an understanding of the future.

At the same time, philosophical analysis must reflect the dialectics of social, natural-biological, and human factors of development in the context of global problems. This significantly influences both the definition of a strategy for resolving these problems and, in general, the forecasting of and research on the future of man and that of the human race. In this regard, scientific philosophy faces many new tasks. As bourgeois and reformist thought has been actively reacting to global problems, it is important to critically evaluate the approaches and methods that it proposes for resolving them and also the new myths and utopia that have been produced.

One should above all note that because they remain unresolved, global problems today emerge primarily as an immediate *threat* to human civilization, and on a scale, moreover, that is unprecedented in history. Never before has man possessed such powerful means of mass destruction and annihilation or such an effective potential for influencing the environment. Never before have there been such serious grounds for concern as to whether the remaining natural resources would be sufficient for the present generation and its children. It is quite clear that all these problems are global in nature and that the wellbeing of both present and future generations and the future of the Earth's civilization itself will depend on whether reasonable ways are found to solve them. The search for future prospects that are favourable to the human race and for new solutions to both old and newly emerging global problems is producing a situation in which the future is beginning to exert an increasingly perceptible influence on the present (the "Oedipus effect"). Turning to the future in the context of global problems, accordingly, has also become a procedure, of sorts, for resolving many current socio-economic and moral-humanistic problems.

In many cases, moreover, attempts to interpret global problems as some general threat to human civilization and to its future serve the objectives of imperialist hegemonism and of efforts to draw attention away from the unquestionable fact that the very emergence of global problems, not to mention their intensification to crisis dimensions, is associated with the capitalist socio-economic formation and that it is precisely the imperialist stage of the latter that has produced a global threat to the human race. It is in the stage of imperialism that the internationalization of the human race's material and cultural life, which had begun and gained intensity during the establishment of the capitalist formation, developed a perverted form and produced many negative consequences. Accordingly, both the globalization of existing problems and the emergence of new problems of joint concern to the entire human race require adequate social structures that can only be provided by peace, socialism and democracy.

Yet in stressing the advantages of socialism in that respect we by no means wish to describe man's future prospects and those of the human race exclusively in optimistic terms. Marxist-Leninist science faces new questions concerning the future of the human race and the prospects of communist development in the context of global problems, and in this regard, optimistic considerations of a general nature inferred from scientific and technological assumptions and forecasts do not suffice.

We know, of course, that scientific and technological progress opens new horizons for the human race: such achievements of science and technology as the use of thermonuclear energy, biotechnologies, putting man in outer space and others, will create a basis for resolving many urgent problems. But it would hardly be reasonable to ignore the difficulties faced by the human race due to the depletion of non-renewable natural resources, a declining energy base, etc. While overcoming these difficulties will require considerable efforts on the part of society as a whole, it will also depend on the ability of each individual to actively participate in solving mankind's global problems.

This is why it is important to have a well-considered scientific strategy that would also take socio-ethical and humanistic premises and factors into account. In this context, an important role is played by efforts of scientists concerned not only with developing theoretical solutions to individual global problems, but also with constructing global models of world development. While the latter often possess substantial shortcomings in terms of the world views and the methodology that they reflect, they nevertheless contribute to the solution of global problems in the spirit of science and humanism. Beyond this, in concentrating the efforts of many sciences, both natural and social, global modeling stimulates cognition and leads to reexamination of a number of problems of human existence, problems related to world outlook.

At present, however, proposals and activities associated directly with the development of concrete scientific and social programmes for solving global problems and perfecting already known methods for implementing strategies and conceptions are more tangible and relevant. As scientific and social thought advances towards the essence of global problems, it is becoming increasingly apparent that solving them is an extremely complex task. For the first time ever, their scope now extends to all countries and all regions. None of the problems, moreover, can be solved separately, independently of the others, or with the help of traditional methods and means tested in past historical practice.

Today it is also generally recognized that fragmentary and uncoordinated actions of individual countries, and even those of groups of states, can no longer overcome global contradictions, and that a joint strategy must be developed for the entire human race regardless of its heterogeneity in terms of classes and social groupings or its economic, political and cultural differences. In the context of peaceful coexistence of states with different social structures, such a strategy requires international cooperation in solving global problems. The result is a still greater internationalization of the human race's activities as its economic and cultural development becomes more internationalized. The internationalization will provide a basis for the further development of human civilization.

Global problems, therefore, are already exerting a substantial influence on all aspects of humanity's life, including its material sphere and culture, its world view and ethics. This influence will increase beyond the year 2000. Mankind's destiny depends on whether we will be able to find, as quickly as possible, ways and methods to at least partly solve these problems or even reduce the intensity of the growing threat.

Yet in defining scientific and social approaches to solving global problems Marxists encounter an active and comprehensive ideological opposition of anti-Marxist bourgeois-reformist forces. Their activities also extend to futurology and to the sphere of global problems. One should not underestimate the negative influence of that propaganda activity, which occasionally turns to lofty philosophic abstractions and humanistic ideas, only in order to redirect them against Marxism and communism. Frequently apocalyptic prophecies of a possible (or even inevitable) end of human civilization and of the human race itself are made to serve such unsavory objectives. References to global problems and to negative social consequences of the scientific and technological revolution also play a significant role in these prophecies.

It would obviously be unfair to ignore the genuine concern of a substantial number of researchers in futurology over the destiny of man and the human race. These are scholars who do not share the positions of Marxism or view socialism and communism as a real alternative to capitalism, but still are seeking a way out of the tragic situation. They approach the problem in terms of humanism and utopian conceptions concerning social and philosophical "mankind's objectives" that would, they believe, lead humanity into the future along some kind of optimal path.

As before, Marxists will continue to criticize the utopian nature of these theories; still they also favour *a dialogue* and cooperation with these scholars on many vital modern problems, including global ones. Marxists advocate an alliance with them in the struggle for peace and disarmament, democracy, man's freedom and dignity, and his future. Marxists consider that in many cases this dialogue offers a reliable path towards a scientific position for those who are genuinely concerned with man's future and that of the human race.

But the above does not apply to the basic world view and the general moral direction of bourgeois and bourgeois-reformist "projections into the future" made in the context of global problems. Such projections are characterized by a "mutual complementarity" of pessimistic and optimistic conceptions and premises. Their latest forms include a mixture of optimism and pessimism in neo-conservative and bourgeois-reformist forecasts. All of these are either directly juxtaposed to Marxism and communism, or else represent attempts to find an approach that is interpreted in terms of the notorious vision of "convergence" of capitalism and socialism, or as a certain alternative movement, whose role is to leave the well-trodden paths of social hopelessness and create new forms of living.[1] In that context, scientific and technological progress is declared to be the source of all kinds of evil, while the meaning of life is found in realiz-

---

[1] Walter Hollstein, *Die Gegengesellschaft: Alternative Lebensformen,* Verlag Neue Gesellschaft, Bonn, 1980; Winfried Schlaffke, *Abseits: Die Alternativen—Irrweg oder neue Weltkultur?* DIV, Cologne, 1979.

ing strivings that are directly associated with the human self.

One can observe major shifts from one extreme to another in the forecasts of the future sponsored by the Club of Rome. Let us compare, for example, the most recent reports made to it with the initial global models of world development produced by Jay Forrester in his *World Dynamics* and Dennis Meadows *et al* in *The Limits to Growth.*[1] It is striking to see the change that took place in the approach to the very same problem during a single decade, that of the 1970s. While the initial report to the Club of Rome emphasized the role of scientific and technological measures in solving global problems in the context of "zero growth" in production and population, the subsequent report by Mihajlo Mesarovic and Eduard Pestel, entitled *Mankind at the Turning Point,*[2] proposed a mode of "organic growth" and added social and ethical proposals to scientific and technological ones. These were addressed both to society as a whole and to individuals. As many Marxist researchers have noted, these proposals were abstract and utopian in many respects, and their authors' position is graphically illustrated by their study's epigraph, namely, "the world has cancer, and the cancer is man" (A. Gregg).

Of course, such a position could hardly be expected to inspire the future generations to whom Mesarovic and Pestel addressed themselves. But it already contained the basis of the shift reflected in subsequent reports to the Club of Rome prepared by Dennis Gabor and his co-authors, and also Ervin Laszlo and his team.[3] That shift led to the recognition (even though in an undifferentiated, abstract form) of the role and significance of

---

[1] J. Forrester, *World Dynamics,* Wright Allen Press, Cambridge, Mass., 1971; Dennis Meadows *et al, The Limits to Growth,* A Report to the Club of Rome. Project on the Predicament of Mankind, Universe Books, New York-London, 1972.

[2] Mihajlo Mesarovic, Eduard Pestel, *Mankind at the Turning Point.* The Second Report to the Club of Rome, E. F. Dutton & Co., New York, 1974.

[3] D. Gabor, U. Colombo, A. King, R. Galli, *Beyond the Age of Waste: Science, Technology and the Management of Natural Resources, Energy, Materials and Food.* A Report to the Club of Rome, Pergamon Press, Oxford, etc., 1978; Ervin Laszlo *et al., Goals for Mankind.* A Report to the Club of Rome on the New Horizons of Global Community, Dutton & Co., New York, 1977.

social factors in solving the human race's global problems. Increasingly evident was an anthropological bias, an emphasis on the need to change man himself, his consciousness, morals, education, etc., as prerequisites and conditions for all other changes, including scientific and technological as well as social changes, that would make it possible to solve the global problems of world development. In many cases the Club of Rome and its leaders (above all the late Aurelio Peccei) reached a conclusion that is, in essence, not unfamiliar in the history of thought, in particular in philosophical anthropology, the conception of humanism of Julian Huxley, etc. That position was clearly evident in both Peccei's *The Human Quality* and in the most recent reports to the Club of Rome.[1]

Peccei also expressed his abstract-humanistic anthropological approach in a lecture entitled "Facing Unprecedented Challenges: Mankind in the Eighties" presented to the International Institute for Applied System Analysis (IIASA) in November 1980. In it he sought to define the future in terms of global problems and listed as necessary measures the adoption of regional and global solutions, regulation of the world as a whole, and, as a prerequisite, self-management of individuals. In Peccei's opinion this requires above all that individuals learn to live in harmony with the real world, meet their responsibilities towards future generations and other forms of life, and develop their own potentialities. This is why, he concludes, global problems should be viewed from the point of view of man's essence, with due consideration of what man can achieve with the help of science, energy generation, and whatever.

Peccei's premise is that the development of modern man is not yet complete and that his creative potential, still not fully utilized, is awaiting its development. If, however, this does not take place, then the currently existing gap between the real world and its problems, on the one hand, and our understanding of that world, on the other, will increase.

One also finds an abstract-humanistic approach to global

---

[1] Aurelio Peccei, *The Human Quality*, Pergamon Press, Oxford, etc., 1977; J. W. Botkin, M. Elmandjra, M. Malitza, *No Limits to Learning.* A Report to the Club of Rome, Pergamon Press, Oxford, 1979; B. Havrylyshyn, *Road Maps to the Future. Towards More Effective Societies.* A Report to the Club of Rome, Pergamon Press, Oxford, 1980.

problems and ways for solving them, an alarmism, and a utopianism in Peccei's *One Hundred Pages for the Future.*[1] It, too, is concerned with man and the human race more generally, their need for universality and universal harmony, and a solidarity that prevails over all political and social frameworks. Peccei again returns to the thought that the human race is threatened by a moral and ecological bankruptcy and its countless consequences. But since man is "the author of his own history", Peccei appeals for urgent measures in solving global problems, and expresses the hope that the human race can emerge from the crisis and build the desired future. In such a context, referring to "our Marxist friends", who stress the importance of modern science and of the revolution in science and technology which plays "a primary role" in the creation of the new society, he alludes arbitrarily to the closeness of that position to the views of "many who are not Marxists", who also believe that science can help resolve nearly all of the human race's problems.

Subsequently Peccei states that there is a great need for generalized and integral knowledge, a synthesis of sciences, and that any progress can be scientific only when it represents, above all "a moral, social or political advancement, an improvement in our customs and behaviour... In a word ... our scientific progress must be cultural above all."[2] In other words, even though this is done in a form that we find in many ways unacceptable, Peccei arrives at ideas that are clearly and distinctly expressed precisely by Marxists, beginning with the classical works of Marx himself. Of course, these ideas differ substantially from the scientistic and technocratic views that Peccei ascribes to Marxism. Turning to Marxism presupposes a serious and fundamental analysis of the Marxist-Leninist conception of the scientific and technological revolution as it relates to man's development as well as a study of the corresponding scientific works that were mentioned earlier. But this does not appear to accord with Peccei's philosophical-anthropological ideas, precepts, and conclusions. In the sections that follow we will see whether and to what extent these ideas are really original.

---

[1] Aurelio Peccei, *One Hundred Pages for the Future,* Pergamon Press, New York, 1981, pp. 21, 89.

[2] *Ibid.,* p. 96.

The same applies to the philosophical-anthropological foundations of the report to the Club of Rome prepared by Bohdan Havrylyshyn under the promising title *Road Maps to the Future*, whose recommendations and conclusions are endorsed in an introduction by the Club of Rome's official leaders, namely, King and Peccei.[1]

That report states a need to establish "a new world order" that would maximally correspond to "human nature".

Then the abstract utopian constructions inferred from an absolutization of "human nature" are transformed into concrete political proposals and conclusions possessing a distinct class orientation. And this, it would appear, is not accidental, even though it has no direct relation to the emphasis on the role and significance of "the human factor" in resolving global problems and the problems of human civilization.

Characteristically, in the activities of the Club of Rome the absolutization of "human nature" is not only accompanied by utopian constructions but also coexists with conceptions of progress expressed in terms of technology. Within the framework of that approach, based on an erroneous world view, certain major processes in the development of contemporary civilization are identified.

In particular, this is clearly evident in the report to the Club of Rome entitled *Microelectronics and Society. For Better or for Worse*[2]. Its aim is to identify difficulties, possibilities and problems and to make it possible for wide social circles to understand what it is that the development of microelectronics is bringing, thus stimulating serious discussions of strategies that may place new technologies at the service of the human race rather than at its destruction. The report touches on many aspects of that complex problem, namely, technological, managerial, political, etc.

Emphasis is placed on the general consequences of introducing microelectronics into industry and everyday life, managerial activities and office work, the sphere of services, on its influence

---

[1] B. Havrylyshyn, *op. cit.*

[2] *Microelectronics and Society. For Better or for Worse.* A Report to the Club of Rome. Ed. by G. Friedrichs and A. Schaff, Pergamon Press, Oxford, New York, 1982.

on politics and international relations, and on the socio-economic development of various countries and regions.

With regard to future prospects for introducing microelectronics, the authors note that if one approaches the world's future from the point of view of the impact of technology, and in particular of microelectronics, on the development of various regions, then one could state that "we are entering a period of deep transition which may last from thirty to fifty years before leading to a completely different type of world society with much greater numbers, changed values, new political and administrative structures, entirely novel forms of institutional behaviour and a technological basis very different from what we are familiar with today, which will influence lifestyles fundamentally in all nations and all cultures."[1] There are three fundamental causes for such a transition: population growth, the need for a more or less comprehensive updating of the planetary system of energy production, and the impact of new technologies, in particular those based on microelectronics.

The authors rightly consider that the deep social changes that result from the wide introduction of microelectronics will call for serious measures in developing skills for managing and regulating these processes. Having made such conclusions, however, they do not proceed to differentiate between existing social systems and establish their respective ability or inability to develop the positive and reduce the negative social consequences of the development of microelectronics. They note that in cultures based on technology there are always tendencies towards a dichotomy between those who understand its nature and functioning and those who simply utilize it without any thought. In the context of wide application of microelectronics that dichotomy may evolve into an opposition between an informed minority and an ignorant majority, and lead to the emergence of a new technocratic elite. An even more serious social consequence may be the isolation and alienation of individuals, of the type that is observed already today in the form of television hypnosis. "The cashless, jobless, information society could easily exaggerate this effect. The concentration of information inlets in a single room, the focus of communication possibilities, many of them impersonal

---

[1] *Ibid.*, p. 20.

and distant, the density of educational and entertainment channels, these and many other factors could immobilise and isolate the family from human outside contact. This could too easily lead to a creeping alienation of the individual, not the active counter-cultural withdrawal we see today, but a passive and insidious alienation with loss of human dignity and self-esteem."[1] This represents a one-sided socio-politically undifferentiated position that identifies socialism with capitalism. The erroneousness of that position becomes particularly evident when the authors discuss the problem of employment associated with the future prospects of microelectronics.

Of considerable interest are numerous conceptions that relate to the new perception of man's essence in the context of high levels of development of science and technology. In particular, this concerns what is referred to as the systemic conception of man in a book by another contributor to the Club of Rome, namely, in Edwin Laszlo's *Le Systemism: Vision Nouvelle du Monde*. This specific perception of the problem of man in the context of the scientific and technological revolution does not provide, it seems to me, a correct answer to the question, for the author seeks to rise above the struggle between socialism and capitalism and to find incentives for further advancement only at the level of scientific knowledge and philosophy.

The same is also clearly evident in *The State of the Planet*,[2] a report on global problems prepared by the International Federation of Institutes of Advanced Studies which is directed by Alexander King. As does a number of latest reports to the Club of Rome, it refers to the inherent external limits and internal constraints in societies and individuals. The limitations are said to be inherent in the very nature of human beings, which possesses numerous qualities that allow man to emerge from among the world of animals and subordinate other species to himself. But today his nature represents the direct opposite of what is required for a harmonious development of human society. Much is said in the book about the importance of better mutual understanding, global solidarity, the need for a new world economic order, etc. Moreover, in the view of the authors a leading role

---

[1] *Microelectronics and Society. For Better or for Worse. Op. cit.* p. 29.

[2] *The State of the Planet.* A Report prepared by the IFIAS. Ed. by A. King, Pergamon Press, Oxford, 1980.

in decision-making will be played not by society's technological and scientific potential, but by the wisdom of individuals, above all leaders. In short, what is presented as a new idea turns out to be the very ancient one concerning the separation of knowledge and wisdom, while an exaggerated emphasis is placed on the role of personality factors in the historical process.

The above one-sided methodological approaches to global problems pertaining to the revolution in science and technology and to ways and means of solving them today and in the future are only specific instances of more general processes in the evolution of bourgeois-reformist thought in the West that are expressed in scientism and anthropologism as opposite forms of studying the role of science in the modern world.

Scientism and technocratism, on the one hand, and anthropologism, on the other, form the two poles of Western futurology. They complement each other, as it were, for they rely on the same type of methodology based on metaphysical absolutization and juxtaposition of individual factors of social development and the isolation of the latter from wider social contexts.

It must be said that today many forecasting projects, especially those sponsored and financed officially, are written in terms of scientism and technocratism. They are prepared for the "establishment" and do not even allude to the possibility of some position other than faith in the omnipotence of science and technology. Such "pragmatized" scientism and technocratism would seem to be especially dangerous, particularly since it plays a predominant role in capitalist societies.

As distinct from explicit scientistic and technocratic conceptions, anthropologism, while not rejecting science and technology in principle, seeks to reduce their negative, "dehumanizing" character. Such a critique of science possesses ancient traditions. Martin Heidegger, for one, sought to strengthen the philosophical prerequisites for liberation from technological determinism, since, in his view, one could envisage a future for technological civilization only in terms of a further dehumanization.[1] Heidegger postulated a certain "turn" by man towards himself, a spontaneous emergence of a new spiritual atmosphere within which

[1] Martin Heidegger, *Über den Humanismus.* V. Klostermann, Frankfort on the Main, 1947.

man would gain a new perspective on both himself and the nature of his relations with everything in existence, including the meaning and significance of technology. Otherwise, Heidegger expects, the rapid development of technology will lead to a catastrophe. This is why one must free oneself from a "technical interpretation" of thought, with the help of intuition combine wisdom and poetry and become conscious of the common ontological basis of both art and technology.

This critique of science was followed up by existentialists and also by the Frankfort school, especially Herbert Marcuse and Jürgen Habermas. In opposing scientistic and technocratic conceptions which assert that the development of science and technology represents the dawn of ideology, the representatives of the Frankfort school see the revolution in science and technology as the source of a transition to a new ideology that is called technical rationality. They, too, absolutize science and technology in their own way, oppose them to other aspects of social reality, and on certain issues hold a sharply anti-scientistic position. In *One-Dimensional Man,* subtitled *Studies in the Ideology of Advanced Industrial Society,* Marcuse writes that "in the medium of technology, culture, politics, and the economy merge into an omnipresent system which swallows up or repulses all alternatives. . . Technological rationality has become political rationality."[1] Accordingly, the claim is made that it is "not the application of technology but technology itself is a form of dominance (over Nature and over individuals) that is methodical, deliberate, and scientifically calculated."[2] In short, according to Marcuse, because it penetrates all spheres of modern "industrial" society, scientific and technological progress creates specific thinking and behaviour characteristic of "one-dimensional" man.

"One-dimensionality", according to Marcuse, thus results from the predominance of "technical rationality" in society, which entails a forfeiture of political and ideological demands, a leveling off of needs, etc. In other words, we are dealing with an absolutized form of scientism and technocratism, although the real-

---

[1] Herbert Marcuse, *One-Dimensional Man. Studies in the Ideology of Advanced Industrial Society,* Routledge & Kegan Paul Ltd., London, 1964, p. XVI.

[2] Jürgen Habermas, *Technik und Wissenschaft als "Ideologie",* Suhrkamp Verlag, Frankfort on the Main, 1968, p. 50.

ity of its absolute predominance in modern capitalist society, where contradictions are mainly defined by relations of property remains an open question. Marcuse, however, lets that basic contradiction, which also points to the real social forces capable of overcoming capitalist social relations, slip by unnoticed. He turns to "human nature", thus abandoning, for all intents and purposes, the sphere of actual social relations. Instead, anthropologism is opposed to scientism and technocratism as their obvious antipode, while man's "new" needs, moreover, are inferred from a Freudian "instinct of life", whose liberation creates a "new sensuality" that is supposedly able to revolutionize modern society.

This version of anthropologism proposes that the dominance of "technological rationality" and the corresponding "one-dimensionality" of man be overcome through some "anthropological revolution". The latter, according to Marcuse, gives birth to a "new man" who creates a "new morality", "new language", "new science", "new technology", and finally, a "new society". Thus, everything is "new" in that conception, proposed by Marcuse, only its utopian and anti-scientific character retains the same old, traditional features of all versions of anthropologism which rely, in criticizing scientism and technocratism, on an abstract conception of "human nature", rather than on specific factors. As a result, solutions to the problem of overcoming the pre-dominance of "technical rationality" and "one-dimensionality" in man turn out to be radical left-wing utopia and "ideologies of rebellion" rather than a realistic goal in social practice.

Habermas, another representative of the Frankfort school, seeks to resolve the contradiction between scientific and technological progress and man's needs by setting the predominance of "technical rationality" in opposition to the hypothetical possibility of unified development of science and technology and human needs, i.e., by achieving—through a number of reforms in the managerial system—their correspondence to the general interests of individuals.[1] That reformist programme, which does not involve any radical social changes as a prerequisite for the "humanization" of science and technology, represents, at best,

[1] See: Jürgen Habermas, *Die Legitimationsprobleme im Spätkapitalismus,* Suhrkamp Verlag, Frankfort on the Main, 1973.

merely still another utopia and a speculative attempt to smooth over the extremes of "technocratic optimism" and scientism.

The "anthropological revolution", proclaimed by Marcuse as a "rebellion against technocracy", did possess some objective foundations in the realities of capitalism, while its anti-scientistic orientation did rely on a critique of actually existing tendencies of scientism and technocratism. But these attempts to arrive at a positive solution of the problem were a far cry from the real needs of society and could serve as inspiration mainly for various extreme left-wing movements with romantic utopian illusions and negativism in relation to science and technology. On the same basis, one witnesses an increasing development of so-called critique of science organized under anti-scientistic slogans, which occasionally turns to apocalyptic prophecies and has adopted a negative attitude towards science and technology as its principle and as a basis for solving man's problems. Advocates of these conceptions capitalize on mythological precepts concerning the assertedly "demonic" nature of science and claim that the latter is capable of threatening man, being as it is a force which is hostile to life and which is increasingly escaping from under man's control and is acquiring a life of its own as it were. Accusations directed at science and technology—that are, it is said, creating only an illusion that man is powerful while in actual fact he has become a slave of industry, that are destroying the environment, producing a degradation of culture and morality, and finally, threatening to destroy the human race itself—usually rejoin conceptions that view any forms of scientific and technological progress as hostile to the development of the "integral personality". In announcing a cult of man and declaring as their ideal the "authentic", unalienated personality, advocates of such conceptions seek to make it appear that there is some other way that provides for both the future of the entire human race and for the comprehensive development of each individual. That way is not associated, in their opinion, with the increasingly rapid progress in science and technology, based on just social relations guided by reason, and transforming all spheres of human life.

To oppose such conceptions does not imply, of course, that one wishes to ignore the fact that scientific and technological progress today is often developing in a one-sided manner and also generating negative phenomena. One cannot ignore the fact, however, that these phenomena can be overcome in the future only with the help of science and technology which develop within social conditions that are worthy of man. It is not scientific and technological progress itself that produces negative phenomena but an insufficiently deep and comprehensive development of science and technology, their distortion and deviation from the humanistic objective that results from the operation of social factors which do not correspond to that objective.

But the romantic "critique of science", which addresses itself more to emotions than to man's reason, does not take this decisive circumstance into consideration. Its call to overcome the "deformations" to which man and his culture have been subjected as a result of scientific and technological progress and industrialization, and to restore man's "multi-dimensionality" and "integral character" as an alternative to his "one-dimensionality", is linked to utopian hopes as regards certain shifts in culture, in consciousness, and in man's morality. A similar attitude is also taken by those thinkers who associate all modern contradictions of scientific and technological progress with the "incompleteness" of Nature and the imperfect character of man.

Such a shift to the individual and to the inner world of the personality represents, to a certain extent, a protest against the anti-human society that represses the personality and levels it, making it "one-dimensional". But also manifest here is a failure to understand the indissoluble links between the individual and society, and the fact that social change is the key to the spiritual liberation and development of the individual.

Such conceptions of man distract from establishing the actual contradictions inherent in the social conditions which govern the functioning of science and which are characterized by class antagonisms. They emphasize certain phenomena that accompany scientific and technological progress, but that are represented as "universal" and "fatal", as the unavoidable fate of the human species.

Such an approach is clearly evident in the works of Theodore Roszak, Professor of History at California State University, and in his theory of "counter-culture". In particular, in *Person/Planet. The Creative Disintegration of Industrial Society* he seeks to show how "the needs of the planet and the needs of the person have become one, and together they have begun to act upon the central institutions of our society with a force that is profoundly subversive, but which carries within it the promise of cultural renewal."[1] The emphasis is placed on the assertion of the sovereign right of the personality to self-discovery, the discovery of a personal meaning in life, and one's personality symbol in facing grief or death. Roszak sees the personality as equally threatened by capitalist self-interest and by "a militant class-consciousness that would politicize the personality out of existence".[2]

Obviously, such an approach cannot be positive, and this also deprives of value the author's attempts to make use of the "personality—planet" relationship as a basic indicator of a healthy economic and ecological policy. Roszak adopts an explicitly anti-scientistic position and places the responsibility for the oppression of the personality on science. He addresses the same accusation to Marxist humanism, since the latter does not break "with the cultural dominance of science or with the imperatives of high industrial integration, and so none can defend our personhood as it needs to be defended".[3]

"A new human identity" requires "a politics that has outgrown the modern infatuation with science and industrial necessity."[4] But the author does not define what this means in reality, and only refers to the need for a "spiritual dimension of life" in the spirit of "the mystical anarchism" of such persons as Lev Tolstoy, Walt Whitman and Mahatma Gandhi. He vehemently attacks Marxism, which, he asserts, while appealing to science, allegedly refuses as a matter of principle to recognize personal freedom, creativity, and transcendental strivings. Roszak concludes that the world needs a new monastic order and a new type of monastic life to help it survive when faced by the prospects of social uncertainty and economic chaos.

So much for the outcome of the anti-scientific utopia and "the adventure of self-discovery through to its planet-saving purpose",[5] that is proposed to the world by the theoretician of

[1] Theodore Roszak, *Person/Planet. The Creative Disintegration of Industrial Society,* Anchor Books, New York, 1979, p. XIX.

[2] *Ibid.*, p. 31.

[3] *Ibid.*, p. 117.

[4] *Ibid.*, p. 119.

[5] *Ibid.*, p. 321.

"counter-culture". One should probably not expect anything more substantial from these romantic and utopian theories which portray the development of science, technology and man in an extremely abstract form. As Marx noted, "it is as ridiculous to long for a return to that original fulness as it is to believe that the present complete emptiness must be permanent. The bourgeois view has never been more than the opposite of that romantic view, and so the romantic will accompany it as a justified opposite till its blessed end."[1]

Marx's anticipations were confirmed by the way in which the radical left-wing ideology of "critics of science", in particular of the "new left", evolved. Marx's conclusion also fully applies to the "new philosophers" and the "new right" in France, made extremely popular in the West by the bourgeois press, radio and television in recent years. The "new philosophers" who gained social prominence during the second half of the 1970s are the successors of the "new left", who had become disappointed in their political methods and concentrated their efforts in the struggle against totalitarianism and authoritarianism on the field of culture. Opposed to humanistic culture in general, the "new philosophers" declare their opposition to the Marxist understanding of humanism, which in their opinion is a "science on dominance". The "new philosophers" assert that man in the traditionally humanistic interpretation no longer exists. They replace him by a Freudian-structuralist model of man whose "self" is, in effect, reduced to sexual desire, the satisfaction of which is said to liberate man.

As for the views of the "new right", who attracted the public attention in the late 1970s, following the publication of the works of such authors as Alain de Benoist, Michel Poniatowski and Louis Pauwels, their conception of man also refers to modern science, in particular to biology, genetics and ethology.[2]

---

[1] K. Marx, *Grundrisse der Kritik der politischen Ökonomie (Rohentwurf, 1857-1858)*, Verlag für fremdsprachige Literatur, Moskau, 1939, p. 80.

[2] Already in 1967, however, *Groupement de Recherche et d'Études pour la Civilisation Européenne* (G.R.E.C.E.) (Group for the Study of European Civilization), formed in France, advocated what the "new right" came to stand for. Pierre Vial, its general secretary (since 1978) has formulated that group's basic ideas in *Pour une Renaissance Culturelle: Le G.R.E.C.E. prend la Parole* (Copernic, Paris, 1979).

They only seek to prove that the decisive role is played by the "principle of distinction" and the need to create a "thinking and active elite", "a new aristocracy" that would allegedly be able to "revive the West" and solve the global problems that threaten it. In *Maiastra,* their manifesto, of sorts (Maiastra is the name of a legendary bird that gave people courage and will at times of trial), the "new right" assert that the main thing is new incentives to live and ideals for which one can die, while everything else is a purely technical problem.[1] They infer their position from a universal "crisis of ideology", including Marxism, with which they wish to achieve a "positive break" by "scientifically" juxtaposing to it their own solutions to basic problems.

The above goal is served by their "new philosophical anthropology". While it does refer to facts drawn from biology, in particular, to the proposition that man's abilities are governed by genetics, it actually misrepresents them in a sociobiologistic spirit. It should be noted that critical pressure had compelled a part of the "new right" to somewhat revise their position. In rejecting accusations of social Darwinism, "biological materialism" and reductionism they assert that man is a being "open to the world" and "unoriented", a being which creates itself. The group claims that it rejects any version of determinism whatsoever, including its biological, racial, economic, sexual, structuralist, or metaphysical versions. The group defines man as the only source of reason within a Universe in which chance is merely another name for necessity. They advocate the "right to difference", a return "to the roots", oppose "economism" (ascribing it to Marxism) and advocate replacing "the mode of life" by a meaning to life.

These declarations contain much criticism of distorted forms of life in capitalist societies. But attempts to extend this criticism to the socialist community are misconceived. And that is where extremes converge: the "new left", through their "new philosophers", are preparing the way for the "new right", who are replacing appeals for "a cultural revolution", designed to cast off scientific rationalism and humanism, as well as man himself, by a programme for establishing a "new aristocracy"

---

[1] *Maiastra: renaissance de l'Occident?*, Plon, Paris, 1979.

and by an open criticism of Marxism and its teaching concerning man.

These mental constructions are anti-humanistic in their very essence, for they return human thought to earlier stages and revive old prejudices.

As we have seen, the implicit philosophical-sociological message of many modern conceptions of man and of his future is either an absolutization of science and of objective knowledge (scientism) or else a cult of abstract man that is opposed to science (anthropologism). Yet this alternative does not always occur in a pure form.

For instance, in Teilhard de Chardin's philosophical anthropology science is viewed not only as a means but also an objective. Teilhard protests against all self-serving forms of utilization of science. He believes that science is still utilized inadequately, namely only as "a new means for providing more easily the same old things", i.e., land and bread. But in doing this "we put Pegasus between the traces. And Pegasus languishes—unless he bolts with the waggon. But the moment will come—it is bound to—when man will be forced by disparity of the equipage to admit that science is not an accessory occupation for him but an essential activity, a natural derivative of the overspill of energy constantly liberated by mechanisation."[1] That is a profound thought, and even though Teilhard does not identify the social essence of science, he clearly shows its humanistic mission. According to Teilhard, the human race is inevitably moving towards "a human era of science, it will be eminently an era of human science. Man, the knowing object, will perceive at last that man, 'the object of knowledge', is the key to the whole science of nature. But man, we should add, is the solution of everything that we can know."[2]

Such a humanistic approach (expressed, from a socio-political point of view, in a very general form) has become a rallying principle and aim of activities of an international organization named *Institut de la Vie,* founded in 1960 at the initiative of Maurice Marois. That institute's goal is "global study of questions of human existence" in order to develop a programme of

---

[1] Pierre Teilhard de Chardin, *op. cit.*, p. 307.

[2] *Ibid.*, p. 308.

action for all mankind. To that end it is proposed that a conception of the "coevolution" of man and his culture with the biosphere be elaborated and that contacts and exchanges of views among scientists and persons of different professions and positions in society be encouraged. Stressing the human dimension of science and technology is declared to be the task of every scientific and technical research project.

The problem of science as it relates to man is also considered in many social-reformist utopias. One example is vividly presented in the widely known book by Alvin Toffler entitled *Future Shock*.[1] The author considers that there is an "impersonal metallic note" in most studies concerning the future and seeks to address primarily its "personal" or human aspect. For Toffler the main problem to be considered concerns the spiritual and psychological state, which he describes as the "illness of change" and "future shock", that results from a continuing acceleration in rhythms of life. It is necessary, in Toffler's view, to change the attitude to the future and better understand the role that it already plays in the present. He proposes what he considers to be a new general theory of social adaptation, namely, the "preparation of individuals for the future".

Toffler describes the consequences of the "super-industrial revolution" in considerable detail. He seeks to establish the limits of man's adaptability to accelerating change and analyzes a strategy that would help man survive under such conditions. He believes that the accelerating rate of scientific, technological and social changes disrupts the human race's biological and chemical stability. This is not necessarily bad: changes are not only a vital requirement of life, they are life itself in the sense that life itself is an adaptation. Ultimately, however, man remains that which he was before, namely, a biological system possessing a limited capacity for change. And when that capacity becomes insufficient there occurs, according to Toffler, a shock from man's encounter with the future. This may result from such factors as psychological stress, excessive stimulation, a "bombardment" of sense organs, information overload, and consequences of stress produced by the need to take decisions.

In Toffler's view, the Gordian knot in the network of causes

---

[1] Alvin Toffler, *Future Shock,* Random House, New York, 1971.

producing changes at an accelerating rate lies in scientific and technological progress. Accordingly, no matter how much individual persons may strive to rearrange their own life rhythms, no matter what psychological crutches may be offered to them, whatever are the changes that may take place in the system of education, society at large cannot free itself from the clutches of scientific and technological progress that is acquiring a frantic pace until there is a shift to a conscious management of scientific and technological progress itself. It is not possible to simply turn off scientific and technological progress, nor should we try to do so. In Toffler's view only romantically inclined simple-minded persons can engage in chatter concerning a return to a "natural state". In raising the issue of taming technology Toffler opposes both technocrats and technophobes. He considers that problems bearing on the development of science and technology cannot be solved as if they were strictly scientific and technological, for these are political problems which concern us more directly than do most of the superficial political issues that we are currently preoccupied with.

Toffler does not offer, however, any scientifically grounded proposition as to which social mechanisms can simultaneously contribute to a further deepening of scientific and technological progress and to the use of reason in directing it for the benefit of man and society. Toffler considers equally obsolete both the system, existing in Western countries and based on economic advantage, and the rigid centralization of state economies that is said to exist in socialist countries. According to Toffler, "what we are seeing is the general crisis of industrialism—a crisis that transcends the differences between capitalism and Soviet-style communism."[1]

The super-industrial society, to which Toffler refers, requires a different structure, but of what kind? He does not provide a clear answer, but from the context of his discussion of the humanization of planning, and the art and politics of futurism, which, in his opinion, is capable of forecasting the future, and in view of his arguments concerning the need for carrying out a revolution in the definition of our social objectives, etc., one

---

[1] Alvin Toffler, *The Eco-spasm. Report,* Bantam Books, Toronto, 1975, p. 3.

can certainly conclude that what Toffler has in mind is still the same capitalist society, only somewhat reformed and more adapted to the rapid changes brought about by scientific and technological progress. As for his appeal to not only overcome the technocratism and replace it by a more humane, and more far-sighted and democratic planning, but also to subordinate the evolution process itself to conscious guidance by man, anticipate the future and plan it—all this remains wishful thinking.

This is fully evident in Toffler's new book entitled *The Third Wave*,[1] in which he takes a global approach to the social and political consequences of the revolution in science and technology. In it he speaks of a new civilization that has emerged in association with the present revolution in science and technology. In this case too, however, we see only a reformist version of the same capitalist society, whose main features are clearly visible in all spheres of the "new civilization" that Toffler analyzes—the "technosphere" (including the energy base, production and distribution), the "sociosphere" (family, the system of education and the dominant type of enterprise) and the "infosphere" (the sphere of information).

Toffler proposes the idea of a "globalist ideology" as an "evolutionary necessity", and "a step closer to a cosmic consciousness that would embrace the heavens as well".[2] He does not believe that within the institutional structure of the "new civilization" that he describes a central role will be played by Universities, that is by science and scientists. He associates a transition to that "new civilization"—"a Third Wave civilization" not with the new revolutionary processes but with "a thousand innovations and collisions at many levels... Much depends on the flexibility and intelligence of today's elites, sub-elites and super-elites."[3]

As we see, both scientistic and anti-scientistic conceptions that appear to offer different solutions in terms of the social and personal aspects of the revolution in science and technology still lead to somewhat similar results. In varying degrees they all prove to be myths resting on a futurological basis. These outwardly dissimilar futurological myths penetrate into literature,

[1] Alvin Toffler, *The Third Wave,* William Morrow & Company, New York, 1980.

[2] *Ibid.*, p. 342.

[3] *Ibid.*, p. 457.

films and television. They stir the mind and tastes, lecture, frighten, raise hopes or simply entertain.

For all their seemingly caleidoscopic variety, the futurological forecasts circulating in the West play one ideological role. There is a specific logic to the way in which some of them at times become more prominent than others and to what is proposed as an alternative to them but in actual fact merely complements them. That logic is determined by objective and subjective factors of social development. In particular, it explains why conceptions of "technocratic optimism" that place the primary emphasis on scientific and technological progress coexist so easily in the West with "disappointment in progress", with an anti-scientistic "criticism of science" and "technological pessimism", and with abstract ideas concerning a "humanization of science". All these constructions serve a common goal, namely, to prove that scientific and technological progress and the intensification of global problems place mankind—regardless of existing social systems—before a common problem of survival.

In this situation scientists concerned with mankind's future face a major humanistic task—to justify, i.e., to explain theoretically, not only science and technology but man himself. Today there is a pressing need for a positive scientific programme determining the tendencies in the development of science and man. To propose such a programme is not a simple matter, and to convince everyone who is concerned with these questions of its authenticity—is still more difficult. But this should not prevent attempts if not to resolve the relevant problems then at least to draw the attention of the scientific community and the public at large to them. Moreover, as we have seen, various scientistic and anthropological approaches, alarmism and utopian optimism cannot provide the foundations for the type of view of the world that this solution requires. Only *humanism in science* and *scientific humanism* can be its genuine basis. This is especially apparent today when analyses of tendencies for the future show an increasing shift in the focus of scientific research in the direction of man; accordingly, this requires an increased emphasis on the human dimension of science and on the role of humanistic values and of socio-ethical factors in its functioning and development.

2. New Social and Human Problems at the New Stage in the Revolution in Science and Technology: the Need for "High Touch" Relations; Man as the Centre, Subject, and Object of Integrated Forms of Interaction among Scientific Disciplines; Scientific Cognition and Humanistic Values; a New Type of Science in the "Age of Man"; Science and Ethics: an Alternative or Interdependence?

The new stage in the revolution in science and technology opens new horizons and also produces social and human problems whose scale and magnitude will greatly influence the future of the human race. It is associated, above all, with qualitative shifts in the development and applications of microelectronics, informatics and biotechnologies, with taming the energy of thermonuclear synthesis and with the exploration and use of outer space. While these lines of scientific research have formed the substance of the present revolution in science and technology from the very beginning, they are currently experiencing revolutionary transformations that are producing *fundamental changes* in both production technologies and many aspects of human existence. It is the sphere of technology that is becoming the decisive factor. In particular, the use of increasingly complex computer systems is not only raising the integrated mechanization and automation of production processes to a new level; it also has the same effect on some forms of mental activity. This is a major characteristic of the present revolution in science and technology. The use of robots (particularly numerically controlled robots capable of adjusting their behaviour autonomously, i.e., to change their own programmes) makes it possible to raise labour productivity to a new order of magnitude. The volume of scientific knowledge is expected to double by the end of the century while information flows are expected to increase by more than thirty times. There will be a corresponding growth in the number of personal computers employed not only in production but also in everyday life and in education where their numbers will increase, according to some estimates, from 250 thousand in 1982 to 2.5 million in 1990. As a result, informatics, which rests on advanced electronic technologies, will become one of the major technologies in the twenty-first century. A number of forecasts suggest that in industrially developed countries microelectronic equipment will become a permanent feature of both production activities and everyday life within the next twenty years.

The development of microelectronics has thus initiated a new stage in the revolution in science and technology that has brought the human civilization into the age of robots and informatics which are transforming not only the productive sphere but modern man's entire life. To a certain extent and for some time the development of biotechnologies proceeded side by side with this process, until their paths crossed at their most important points, confronting the human race with still other prospects and technological possibilities. But we also find similar interactions at early stages in the development of computers. In particular, the very emergence of cybernetics resulted from a heuristic application of the analogy between man and machines, while studies of numerous physiological mechanisms governing man's mental activities have provided a basis in the design of increasingly advanced generations of computers, including robots and efforts to create an "artificial intellect". Still, it is only in the last decade that a genuine *biologization* of production technologies has begun, combined with the use of microelectronics in new generations of computers. It is much more, moreover, a future prospect made possible by the most recent revolution in the science of life that began with the major advances in physical-chemical and molecular genetics that led to the emergence of genetic engineering.

In the present context there is no need to present a detailed review of the implications of the above, for the relevant facts concerning the biotechnological applications of genetic engineering are widely known.[1] I am referring not only to the applications of genetic engineering in medicine, but also to agricultural applications that will produce a new "green revolution" and will eventually radically restructure the technologies of food production, and to industrial applications involving the use of biological sources for the production of energy. Subesequently

---

[1] *From Biology to Biotechnology,* the Moravian Museum, Brno, 1981; Luc Joassin, "Les perspectives de la biotechnologie dans ses differents domaines d'application", in: *Problèmes Économiques,* No. 1.773, 12 May 1982, Paris; *Biotekhnologiya* (Biotechnology). Ed. by A. A. Bayev, Nauka, Moscow, 1984.

during this stage in the revolution in science and technology, when science will truly have entered the "age of biology", new discoveries at the interface of microelectronics and genetic engineering become possible; many scientists believe that these will lead to the emergence of bioelectronics as a new technology. Aside from placing life itself into human hands this will make possible the development of new "mixed" systems practicing forms of human reasoning.

What will the implications of this be for both man and society, and what corresponding urgent social and human problems are already emerging today? Like problems relating to the effectiveness and productivity of new technologies and of new principles in organizing and managing production, based on a growing reliance on microelectronics, informatics and biotechnologiest, these are extremely important issues. That is due to the fact that the new technologies do not exist in a social vacuum and do not determine by themselves the total multi-faceted spectrum of human existence that is embodied in particular types of civilization and of human culture. In turn, their own development, too, largely depends on social and human factors, even though it does possess a measure of autonomy which is, as we have seen, occasionally absolutized in various technocratic conceptions. This is why it is so important to turn to the study of social problems that arise at the new stage in the revolution in science and technology, and especially of those that are associated with the development of the new technologies.

If one were to define the main social problem of the new stage in the scientific and technological revolution that follows directly from the requirements of what is known as "high" technology a short answer would be: a need for "high touch" relations. This terminology, borrowed from the lexicon of computer technology, thus acquires a wider meaning. It implies that *increased technological levels in both production and all other human activities must be matched by corresponding increases in levels of social and human development in their interaction with Nature,* by a new civilization and a new humanistic culture deriving from the fact that man is an end in itself of social development (Marx).

Marxist-Leninist theory has long arrived at that conclusion, having established the deep mutual interconnection and reciprocal influence between scientific and technological progress and social transformations, an interrelationship which also includes man's own development and that of his culture, including his relation to Nature. What are the new elements that thus emerge in connection with the new stage in the revolution in science and technology? By raising the intensity of the current problems to its very limit, what it calls for are precisely relations of the "high touch" type between the new technologies, on the one hand, and society, man and Nature, on the other. Moreover, this is no longer simply a vital need, but also a necessary *condition* for both the effective application of these technologies and the very existence of society, man and Nature.

Such a formulation of the problem points to concrete social measures and also defines a context for the present struggle of ideas. It therefore possesses a practical political and scientific meaning. Let us consider how this is neglected in different social systems and also the relevant scientific, semi-scientific, and sometimes, unfortunately, anti-scientific conceptions that have been proposed in this connection in various parts of the world. In turning to these discussions, we will attempt to outline, if only in a general way, the contours of the "high touch" that, in my view, is already required for technology and will be especially needed in the future, above all as the new technologies bear on forms of social organization at the micro and macro levels and on social prospects.

If one only considers the *technological* aspect of the matter, the situation is more or less comparable under socialism and under capitalism. In recent decades developed capitalist countries, especially Japan and the United States, have linked their future to the development of new technologies. The production of industrial robots, and more recently of products of biotechnologies, has been growing especially rapidly. The momentum of evolution of new equipment and technologies has been particularly remarkable if one considers that the manufacture of robots began to develop rapidly only in the late 1960s and early 1970s, following the appearance of microprocessors and of microcomputers. While the first microprocessor was developed in 1971 there were already 250 million microprocessor-based systems and devices in the world in 1980.

As major efforts to develop microelectronics, informatics and biotechnologies are being undertaken in the world, capital investments in these sectors are growing. According to OECD data, investments in computer technology and informatics in developed capitalist countries are of the order of 15 per cent of their GNP. In 1981 the volume of sales of the output of electronic industries was $100 billion, and is expected to quadruple by 1990.[1] The production of robots is developing especially rapidly, and it would appear that Japan will continue to hold a leading position in this field at least until the turn of the century. A number of forecasts suggest that the world market for microcomputers will increase at a yearly rate of 50 per cent (34 per cent in value terms). Capital investments in biotechnological research and applications are also growing at an extremely rapid pace. Data published by the Japanese Association for Industrial Robots indicate that there were 20,268 advanced robots in the world in 1979, of which 14,000 (69 per cent) operated in Japan, and 3,255 (16 per cent) in the United States. There were 6,300 robots of this category in other Western countries. The present author was informed by the Japan Productivity Centre that of 100,000 robots in the world today 70 per cent were produced in Japan and 20 per cent in the United States. It is expected that the rate of growth will increase by an average of 40-50 per cent per year in Japan, and 35 per cent per year in the United States. Accordingly, by 1990 there should be approximately 560,000 robots in Japan (5 times more than at present) and one million by the end of the twentieth century. As for the United States, it is expected that by 2000-2010 the number of its industrial robots will exceed 100,000, and that most of them will be of the "fifth generation". This will make it possible to replace 11 million workers. Other estimates indicate that in 1983 there were 13,000 industrial robots in Japan, 6,250 in the United States, 3,500 in the Federal Republic of Germany, 1,300 in Sweden, 950 in France, 700 in Italy, 350 in Belgium, 300 in Australia and 120 in Finland. Because the definition of what constitutes an industrial robot varies from country to country estimates differ somewhat. But whatever the differences, the use of new technologies and especially of industrial robots leads

[1] Colin Norman, "Microelectronics and the World Economy", in: *Transatlantic Perspectives,* No. 4, January 1981, p. 19.

to substantial increases in labour productivity and in national income; thus, it has become possible to automate production, control output quality, and improve labour conditions—especially in health-hazardous spheres.

According to the standard classification adopted in Japan in 1979 robots are divided into five groups: (1) manually controlled manipulators, i.e., robots guided directly by human operators; (2) robots that perform a sequence of operations defined by particular tasks, conditions and positions; (3) "learning" robots possessing a memory unit and performing operations in accordance with the information that it contains; (4) numerically controlled robots performing preset operations on the basis of mathematically processed information; and (5) "intelligent" robots that adjust their behaviour autonomously. Of course continuing advances in robots, particularly of the last group, largely result from an intensive development of microelectronics and corresponding increases in the information capacities of computers. Today 40 per cent of the latter have memories based on integrated circuits containing 16,000 cells, but Japanese firms have already begun to produce silicon chips containing 64,000 cells. Early in 1980, moreover, they have demonstrated a version containing 265,000 cells.[1] A number of researchers therefore expect that by the early twenty-first century "intelligent" robots will already be endowed with "sight" and "hearing" and will themselves be able to alter their programmes. Experimental versions of such robots already exist in Japan.

These, then, are some dimensions of the "electronic revolution" and of its technological prospects on the eve of the twenty-first century. But what will its "social interface" be, and what problems will urgently require solutions?

There are numerous technological studies of the way in which microelectronics and associated technologies influence production processes, the organization of labour and other factors. A major feature of these technologies is their practically universal character—ranging from the smelting of steel, mining operations, and the exploration of oceans and of outer space to banking and office work, production management, and research in highly complex areas, including biomedical studies.

[1] Colin Norman, "Microelectronics and the World Economy", *op. cit.*, p. 19.

The shrinking size, growing efficiency and increasing flexibility of electronic equipment have greatly reduced the cost of storing and processing information. Telematics allows information to be exchanged not only between persons as well as persons and machines, but also between machines. The future creation of global integrated information and communications systems will also make it possible to achieve a substantial integration of production processes and social processes (such as control over the rational consumption of energy, or telemedicine). This has led some Western researchers and futurologists to assume that industrial societies of the capitalist type will therefore be transformed into "information societies" and that the latter will somehow resolve the basic problems and contradictions currently disrupting the capitalist world despite advances in production that are largely stimulated by the large-scale use of new technologies. This, it is said, is the way in which the social "high touch" will be achieved.

But is this actually the case, and what does high technology actually bring under capitalist conditions? Above all, of course, a decline in the norm of capital accumulation and rates of economic growth, as well as a need to restructure economies. But an issue of particular importance is its impact on unemployment. The new technologies' high labour-saving potential raises labour productivity and eliminates many types of work and of technological operations. Under capitalist conditions this will produce a further increase in unemployment.

In their book entitled *The Old World and the New Technologies,* Michel Godet and Olivier Ruyssen estimate that between 1970 and 1977 production in the Federal Republic of Germany increased by 12 per cent and productivity by 43.5 per cent, whereas employment declined by 23 per cent.[1] They believe, moreover, that while under low rates of economic growth in the European Economic Community the new technologies do lead to greater unemployment, without that technology unemployment would even be larger. Let us note in passing that more than 1.5 million persons are employed in the electronics industry in the United States, and the number is expected to increase to 2.5

[1] See: M. Godet and O. Ruyssen, *The Old World and the New Technologies,* Office for Official Publications of the European Communities, Brussels, 1981, p. 111.

million by 1987. Yet this is not a net increase, but only partial compensation for the decline in the number of jobs that will follow the introduction of microelectronics in various sectors. By the year 2000 one-third of all workers in the United States will be employed in sectors concerned with microcomputers and computer technology. By the same year, robots may displace up to 75 per cent of the present work force in developed capitalist countries. Godet and Ruyssen also expect that during the next two decades modern industrial (capitalist) societies will become "information societies". The wide use of telematics and microprocessor technology will reduce the number of available positions, causing the socio-economic situation in countries of the European Economic Community to deteriorate and possibly reducing their ability to compete with the United States and Japan. Nevertheless the authors recommend that the Old World accord priority to the new technologies, including biotechnologies. In emphasizing the uncertainties of the geopolitical, energy, scientific and technological dimensions of the world's future course of development they do not preclude alternative possibilities and call on Western Europe to accept severe hardships, define a new development model that would recognize the diversity of possibilities that are implicit in a wide use of microprocessors, and establish all-European firms that could pool their capacities on world markets.

It is, of course, unlikely that an approach that only considers measures within existing capitalist structures would meet the need for social relations of the "high touch" type. More generally, it fails to match the complexity of the socio-economic problems that result from a wide reliance on the new production technologies which influence the sphere of production relations as well as the entire system of socio-economic linkages at the macro level. This explains both the optimistic and also the pessimistic conclusions of numerous Western forecasts, including those produced by technocratic futurology. With regard to the social problems created by the new technologies some Western authors express the view that in the context of free markets rapid technological change imposes excessive economic, social, and personal costs on those parts of society that are least able to assume them. In Great Britain, for example, the wide introduction of microelectronic equipment may reduce the number of available jobs by 5 million, while in the Federal Republic of Germany it may eliminate 2 million jobs in the sphere of office employment alone. The obvious result would be mass unemployment. Yet purely technological measures are proposed in seeking to avoid such developments, such as the creation of an "alternative information system" as an element of a national and international strategy of the workers' movement within the framework of trade union activities. When stated in such a general form, that proposition does not raise any objections. That, however, is only true as applied to its general terms, for in their *practical* impact the proposals do not extend beyond social-reformist measures and fail to solve the fundamental and basic problem of unemployment.

Meanwhile, growing mass unemployment in industrially developed Western countries appears as the *major social consequence* of the introduction of the new technologies. This has led to a search for solutions to that problem, many of which remain within the framework of capitalist relations. Proposals include the introduction of shift work schedules, an end to systematic overtime work, streamlining the training of skilled labour, adjustments in wage rates, and a number of ways to reduce the average working time per year, either by increasing leaves or the number of days off the job, or else by reducing the length of the working day. Occasionally authors associate such a set of measures with the development of a "new industrial civilization", as has been done, for example, by the French economist Michel Guillaume, in *Sharing Work: Another Industrial Civilization*?[1]. In this book, the author presents a reformist utopia based on the conviction that the social problems brought about by the introduction of new technologies can allegedly be resolved in a "new industrial civilization" within the framework of capitalism—by such measures as making labour more meaningful, abandoning narrow specialization, providing a "multiplicity" of productive activities for each person, increasing the mobility of production factors, including that of manpower, and raising social mobility to a high level. But will such a new industrial civilization successfully overcome the crises in employment, in

[1] Michel Guillaume, *Partager le Travail: une autre Civilisation Industrielle?*, Presses Universitaires de France, Paris, 1983.

labour processes and in organization? Will it solve problems of international division of labour, efficiency as well as technological and energy problems? These questions, in the author's opinion, are difficult to answer. Nor is this surprising, for his study does not even come close to a positive scientific solution of these issues.

Other recent conceptions of social-reformist futurology, relying increasingly on the latest advances in microelectronics, informatics, and biotechnology also offer palliatives, rather than genuine solutions. These ideas are emphasized by Alvin Toffler in his *The Third Wave,* in which he seeks to attribute capitalism's current crisis to the decline of the industrial society and its transition to a "new civilization". A similar position concerning a future "information society" is presented by Jean-Jacques Servan-Schreiber in a futurological study entitled *Le Défi Mondial.*[1] Also found in the same group is the Club of Rome's most recent report entitled *Microelectronics and Society. For Better or for Worse.* It should be noted that many such conceptions do address real problems and show an understanding of the need for a "high touch" between the new technologies and social structures. Yet the corresponding conclusions are social-reformist at best, made as they are in the spirit of the "new civilization" concepts. Of course they do differ in many respects from an open apology of capitalism, and to that extent the utopias of a "new civilization" and an "information society" do reflect, even if in a distorted form, some of the real choices open to man's civilization in the future. But they also play an important negative ideological and political role in the context of the continuing active opposition between capitalism and socialism, for in addressing themselves to the new technologies and their influence on social development they seek to define a social approach to the future that would exclude the socialist and communist perspective. But it is precisely there that they guide social thought away from the genuine high level of social responsiveness that the technologies require, both today and in the future. They do indeed require a *new* society and a *new* civilization to replace capitalism, namely, the civilization of communism, in relation to which socialism is a formative stage.

---

[1] J. J. Servan-Schreiber, *Le Défi Mondial,* Paris, 1980.

The socialist society fully recognizes the revolutionary potential, social consequences and technological possibilities of microelectronics, informatics, and biotechnologies. It not only provides a greater scope for their actual application but in its very nature corresponds most fully to these new technologies which require the involvement of vast social funds rather than private capital, even though at a particular historical stage private capital may exceed the material resources available to socialism. The social system of the socialist society allows for making maximal use of the new technologies' basic labour-saving potential without the increased unemployment that unavoidably follows their introduction under capitalism.

That is socialism's principal advantage and the fundamental basis for its high social responsiveness to the new technologies. And it is on that basis that socialist countries have initiated major efforts in recent years to intensify social production by introducing in it the achievements of scientific and technological progress, primarily microelectronics, informatics, and biotechnologies. In particular, more than 19,500 automatic manipulators have been produced in the Soviet Union since the beginning of the eleventh five-year plan period, while the production of microprocessors has increased nearly five-fold and that of microcomputers has more than doubled. Approximately 1,200 automated production management systems have been established by 1983, almost as many as during the entire tenth five-year plan period. In 1983 alone, the above measures have made it possible to save the labour of more than 600,000 persons. In the German Democratic Republic measures that include the use of industrial robots replaced 300,000 work places between 1981 and 1985. During the same period, the production of robots was scheduled to increase from 13,000 to 40-45,000.[1] The GDR has been making major capital expenditures for the development of microelectronics since the second half of the 1970s (nearly one billion marks in 1980 and 2.5 billion in 1984). Major steps to promote the development of new technologies, especially microelectronics

---

[1] Heinz-Dieter Haustein, Harry Maier, "Flexible Automatisierung—Kernprozes der revolutionären Veränderung der Productivkräfte in den achtziger und neunziger Jahren", in: *Wirtschaftswissenschaft,* Berlin, No. 5, 1982, p. 681.

and robotics, have also been taken in Bulgaria. Under its next Five-Year Plan, approximately 70-75 per cent of capital investments in the material sphere will be channeled to fundamentally new technologies and only 25-30 per cent to the partial reconstruction and modernization of currently existing technologies. There is a similar emphasis on developing and introducing new technologies in other socialist countries.

Still another advantage of socialism lies in its capacity to establish large-scale cooperation among many countries with different levels of scientific and technological development in producing new technologies. While there is an intense competition in the capitalist world among Japanese monopolies producing the new technologies, monopolies in the United States, and, increasingly, the monopolies in an "integrated" Western Europe, countries of the socialist community are relying on mutual assistance and close cooperation in developing science and technology and introducing their advances into social production. This is reflected in the further development and deepening of economic, scientific and technological cooperation among the CMEA member-countries which have adopted, on the basis of their national programmes, a Comprehensive Programme of scientific and technical progress in the CMEA countries up to the year 2000. The integrated programme will be the basis for the elaboration of a common integrated policy in science and technology. Particular emphasis is attached to electronics, microprocessors, robotics and biotechnologies. Existing advances in biotechnologies are already impressive and an even more rapid pace is anticipated in the future as they are applied to medicine, agriculture and industry.

While socialism provides major opportunities for solving these problems their solution does not, of course, take place automatically. It requires persistent efforts, creative approaches and an ability to overcome the inertia and conservatism that occasionally hold back manifestations of what is known as "the subjective factor". In the context of the latter, however, our attention is drawn to a more general problem concerning the development of man proper. The "high touch" of the new technologies presupposes the attainment of a high cultural level and an advanced stage in achieving a release of man's essential and creative forces in integral, harmonious forms. A "human responsiveness" of the new technologies and their humanistic evaluation in the light of the possibilities offered for an integral, harmonious development of the essential forces of man as a creative being fully coincide with the new society's basic value orientation, namely, concern for the human being and for his developing cultural needs. Today this fundamental orientation of the theory of scientific communism is being translated into the sphere of practical political activities of the socialist countries envisaged for the period up to until the year 2000 and beyond.

Under socialism men are not "displaced" from production so as to join the ranks of the unemployed as a "redundant" element. On the contrary, they have growing opportunities to engage in production and to genuinely subordinate it to man's material and spiritual needs, as well as to develop their abilities outside the production sphere, which is increasingly assigned to machines. But it would be empty doctrinairism and utopianism to view the actual situation in this regard exclusively in terms of that ideal. One must not ignore the problems and difficulties that the CPSU has been seeking to overcome in recent years by taking concrete measures to solve the social problems arising at the current stage in the revolution in science and technology. This also applies to the socio-economic, organizational, managerial and educational problems associated with the introduction of new technologies and to related issues in the sphere of culture, as well as in general and political education. While Soviet experience in all these respects is already considerable, it does not always match the scale and complexity of the new problems that the socialist society confronts at the new stage in the revolution in science and technology. These challenges follow directly from the proposition that one should combine *in practice* rather than simply in words the advantages of socialism with scientific and technological advances that promise radical technological changes in many spheres of production. At the present time there is considerable emphasis in Soviet scientific publications and in the press not only on successes in realizing that task in Moscow, Leningrad, the Ukraine and elsewhere in this country but also on shortcomings due to poor organization in the production and introduction of robots, their insufficiently effective technological application, the conservatism of individual managers and a lack of personnel trained to operate them. This often greatly reduces the production efficiency of the frequently expensive new technologies and their human significance is often reduced.

A separate major problem whose solution will make it possible to derive the fullest benefit from the advantages of socialism concerns the need for more efficient management in that field thus combining basic and applied research in new technologies with pilot production activities, their mass production, and subsequent introduction in various areas, with the training of corresponding personnel and continuous education and retraining of wide masses of the population designed to encourage the development of new norms of behaviour and a new understanding of the very meaning of human existence in the age of microelectronics, informatics, "intelligent" robots and biotechnologies. At present, unfortunately, the "weakest link" in the new technological processes is often man himself, and this draws our attention, at the new level of scientific understanding of man's problems, to the need for multi-disciplinary studies of man (in his social, moral, medical, genetic, psychophysiological, ecological and other aspects) and for corresponding forms of human development.

A number of measures are being taken in the Soviet Union in developing a reliable basis for solving the whole range of social and human problems faced at the new stage in the revolution in science and technology. These measures include the development of new technologies, improving the national economic mechanism and social forms of the participation of workers in the development of self-government, developing mass initiatives, carrying out the reform in general and vocational secondary schools, and the corresponding work of the USSR Academy of Sciences and of other interested organizations and departments. But time does not wait, and much of what we are doing today and plan to do tomorrow should have already been done yesterday. The rapid development of the new technologies imposes a need for fundamentally new approaches and for considering long-term prospects as well as for using what is already available. There is an awareness of this in many developed capitalist countries as well, for instance, in Japan, where one-third of all expenditures on the national programme to develop the fifth-generation computer is spent on education and the training and retraining of personnel.

Aside from using its many advantages, including those in the sphere of education, in developing new technologies, socialism is also able to mobilize moral as well as material incentives and forms of conscious organization and initiative involving wide masses of the population. This is especially important in encouraging the participation of young people in mastering, developing and introducing equally young technologies.

The "Basic Guidelines for the Reform of General and Vocational Secondary Schools" (1984) place particular emphasis on instruction in computer-related subjects and skills and on the broad use of computers in education. The USSR Academy of Sciences, too, is paying much attention to the development of personal computers and to their use in schools.

Comprehensive support in socialist countries is extended to initiatives of the youth movement in sponsoring the effort to master and develop new technologies. In the Soviet Union a resolution was adopted in 1984 on the participation of members of the Young Communist League and of young people in developing computer technology and studying the fundamentals of its application. Its aim is to involve young people in the development of integrated automation, computerization, microcomputers and personal computers and in mastering their use in production and everyday life. A similar initiative in Bulgaria has produced a comprehensive stage-by-stage programme under which the Young Communist League sponsors instruction in computer operation and the application of computers in production automation, services and management. It is scheduled to be implemented in stages, the first one ending in 1985 and the subsequent ones covering the period from 1986 through 1990.

We do not wish, however, to take a simplistic view of the problems faced by socialist society. It is clear that everything should be done to bring about *fundamental changes* in the human perceptions of and approaches to the new technology; a major role in this regard will have to be played by schools, adult education and by the mass media, including the press. Universal instruction in the fundamentals of the new technologies, and especially of computer technology, is a major social task at the current stage in the revolution in science and technology and a necessary condition for combining its achievements with the advantages of socialism. This will apparently continue to be a major task in the field of science and technology in socialist society around the turn of the century.

It must be noted that the real social and human tasks with which socialism is concerned possess global aspects as well. Throughout the world the new technologies' *coadaptation* with man's development as a personality is producing most acute social and human tensions. These problems are associated with man's need for understanding and using the new possibilities without losing the meaning of human existence in a world in which robots are increasingly "displacing" man from direct participation in production—and not only from routine and arduous forms of work but also from other activities that machines simply perform better. In the West this has occasionally produced sociophilosophical mental experiments that surpass even the vivid and revealing fantasy of Karel Čapek—the original author of the concept of a "robot" in his drama *R. U. R. The Rassum's Universal Robots*. These include the anti-utopias of Aldous Huxley and George Orwell, as well as the most recent myths, whose aim is either to scare people—who are already confused and alienated from society and who fail to understand, in particular, the meaning and significance of the new technologies—or else to instill excessive hopes and false illusions limited to the sphere of pure consumption. Some predictions made by Marshall McLuhan, for example, refer to "the demise of humanism", while the mass media are described as factors that produce deep changes in human nature itself. The media and technologies are seen as physical and material embodiments of culture, and ideologies as merely a decoration or an official dress, as it were, worn by the media in Computerland. Even the possibility of changing human physiology in its organic reality through the media is being discussed. The neoeugenic ideas of a "fabricated man" and even of "biocyborgs" are being revived in new forms based on the abstract possibilities of microelectronics and genetic engineering. But there is also a growing awareness—often, as has already been noted, it is reflected in abstract and utopian forms—of the need for a high level of social responsiveness of the new technologies to man himself and to a new type of humanistic values that capitalism does not and cannot provide but that constitute the essential foundations of communist civilization. Again in Japan there is a constant emphasis on the significance of socio-psychological and cultural-ethical problems associated with the process of robotization. And while in its narrow meaning the conception of "high responsiveness" ("high touch") is employed to refer to "man-machine" interactions of a type in which computers as "sensitive" machines place many "demands" on man, including the renouncing of a number of habits (such as smoking) that reduce the degree of their responsiveness, it also refers to such vital phenomena as growth in professional knowledge, man's general culture, and his moral qualities. This became evident, for instance, during this author's visits to the Sony enterprises, Japan's postal and telegraph services, and Japan Productivity Centre. It was also the subject of repeated discussions at the *Institut de la Vie*'s symposium on coevolution in the age of robots, held in Paris in 1984. In his book entitled *Advanced Technology and the Japanese Contribution,* Moritani Manasori, a prominent researcher studying scientific and technological progress in Japan, emphasizes that in the long-term, industrialism and the technological development on which it depends are determined by the human factor. He also believes that a major characteristic of future technological progress will be its penetration into the world of human feelings and abilities. Music, painting, culinary arts, leisure time, hobbies, sports and education will become major spheres of application of the new technologies and new products. This may be described, in his view, as the technology of the future culture. In his opinion, this combination of culture with technology will determine new directions of growth for which, incidentally, Japan's technological climate and cultural traditions are perfectly conducive.

In a number of other developed capitalist countries as well, long-term plans that include the development of new technologies place an increasing emphasis on their "human touch". In one of its studies France's *Le Commissariat Général du Plan* states explicitly that there is a lagging of social structures, culture and morality behind the changes that result from the development of technology. Technologies are said to be employed in extremely petty-bourgeois ways, above all for advancing further the "consumer society". Many perceptive Western scientists are becoming increasingly convinced that existing economic models cannot encompass the full complexity of the new situation, and that models of a new type must include not only a non-material information sector but also other non-material aspects of human activities as well their environmental interactions.

For instance, Jacques Robin, a French scholar studying advanced systems and technologies as they relate to culture expresses the conviction that from a socio-historical point of view the current shift towards greater use of information technologies is something more important than merely a "third industrial revolution". What is taking place, he believes, is a transition from the "age of energy" to an "information age". In his opinion, the Western countries should develop a "cultural project" to define a "strategy of coevolution" of technology and culture whose aim would be to hold back forms of technological progress that are socially undesirable.

Elaborating on this, he notes that the use of information technologies to subordinate economies to the developing and changing socio-cultural needs of society will require that traditional forms of economic organization and of monetary relations be replaced by new, still unfamiliar forms.

The conclusions made in this regard by André Piatier, who is head of a major research centre in France concerned with the influence of technologies on economies, are also worthy of note. In an article entitled "Innovation, Information and Long-Term Growth"[1] he lists the factors that must be taken into account in a more general evaluation of the functioning of the information sector and in the elaboration of a model designed to increase its effectiveness. He views human capital in a common category with material resources and material production capital, and takes into account such factors as manual labour, intellectual labour, ideas, materialized ideas and progress.

In the author's opinion, the new stage in scientific and technological progress (the emergence of computers, robots, etc.) provides a new basis for increased efficiency. In such a system economic development would not necessarily end in stagnation. Piatier believes that pessimistic conclusions concerning the new stage in the revolution in science and technology can only derive from an excessively narrow understanding of the conception of the social system within which science and technology progress.

---

[1] André Piatier, "Innovation, Information and Long-Term Growth", in: *Futures*, Vol. 13, No. 5, 1981.

Major new prospects, in his opinion, would result from greater decentralization made possible by the development of telecommunications, computerization and informatics. The introduction of numerous innovations, stimulated by the development of informatics, will also raise efficiency in spheres where it is currently declining. The progress in technology that is currently taking place will be followed by economic progress that, in turn, will lead to social progress.

It is true, however, that the author does not describe how this interaction will actually take place, and this factor, naturally, reduces the value of his position in its entirety. It should be noted, nevertheless, that the author shares the views—symptomatic if illusory—that are based on the theory of "human capital" which in its present form appeared in the West in the 1960s. In the opinion of its advocates, that theory can help reprient economic thought. In particular, Lester Thurow, an American economist, considers that "human capital" includes the productive abilities, potentials and knowledge of individuals.[1] Investing in man is therefore recognized as a major issue in the age of new technologies.

While the above approach, which replaces political and economic problems by ideological ones, is unlikely to be constructive, it does focus attention on extra-economic factors whose importance is enhanced in the "age of robots". Marx stressed in *Capital* that in any capitalist machine production even facilitating labour becomes a source of torture, for machines free labour of any content rather than liberate man from labour. For this reason, a growing number of researchers today, including non-Marxists, note that the spread of robots produces still another fundamental humanitarian problem for workers, namely, that of the reproduction of human functions outside the production process.

But this is a social rather than a technological problem that calls for social solutions deriving from a new humanistic vision. Some interesting ideas on this subject are contained in an article by A. B. Cherns, a British sociologist, entitled "Speculations on

---

[1] Lester Thurow, *Investment in Human Capital,* Wadsworth Publishing Company, Belmont, California, 1970, p. 1.

the Social Effects of New Microelectronics Technology".[1] In his view, the consequences of the introduction of microelectronics for employment are more or less clear, even though opinions concerning their scale and depth vary. Some researchers believe that the decline in the labour force will soon assume catastrophic proportions. Others, who are less pessimistic, note that similar fears were voiced following the appearance of computers but were not confirmed. Cherns notes that in order to evaluate and fully utilize the new possibilities offered by microprocessors one must realize what type of society we intend to create by using this technology and what its implications will be. The uses of new equipment should depend on a deliberate choice rather than on extraneous circumstances as has been the case in the past. It now seems quite probable that different societies will use microprocessors—which can potentially be employed for a variety of purposes—in different ways. Some societies, the author concludes, will use them in developing humanistic values, while others may apply them for establishing the predominance of institutions over societies. One can only agree with this view, with the proviso that it is only in socialist and communist society that the conditions needed to place new technologies at the service of man and of his cultural values will be created.

This is the conclusion to which one is led by Marxist methodology which is embodied politically in the struggle of Communist and workers' parties in developed capitalist countries to protect workers' rights as new technologies are introduced. That struggle is not directed against new technology per se but specifically against capitalist forms of its use. Its objectives include a humanistic orientation of scientific and technological progress and its application in production as well as the further development of man, his education and culture. But this requires deep social transformations in the direction of socialism and communism that can also resolve the humanistic problems of science which have become much more pronounced due to the wide introduction of new technologies, including microelectronics, informatics and biotechnologies.

---

[1] See: A. B. Cherns, "Speculations on the Social Effects of New Microelectronics Technology", in: *International Labour Review*, Geneva, Vol. 119, No. 6, pp. 705-721.

It should be emphasized that scientific cognition is a human activity; it can only be carried out by men, and for this purpose men organize into large groups of researchers and use the most complicated technology. This is self-evident and leads to a number of inferences pertaining to the substance of the subject under discussion.

The most immediate one is the conclusion that knowledge generally, and scientific knowledge in particular, can only be obtained through means and in forms that are determined by man's constitution and his psychological and physical characteristics, such as the scope of his memory and the structure and resolution of his sensory organs. Undoubtedly, individuals do differ in terms of these characteristics and abilities, just as there is no doubt that in specific—and, moreover, very wide—limits these characteristics and abilities can be developed through education and exercise. There is also a variety of technical means in existence and under development using which man can extend the range of his cognitive possibilities. Nevertheless, in the process of developing knowledge the subject of cognition is "projected" onto a specifically human cognition mechanism, and must therefore be commensurate with man. That constitutional characteristic of scientific knowledge, as the above is known in traditional philosophy, is linked dialectically to the outward orientation of that knowledge, that is towards a content that exists objectively, independently of the cognizing subject. But this means that in terms of its underlying foundations and origins knowledge cannot be external in relation to man. Externality, incommensurability and extraneousness of knowledge must therefore be attributed to the specific characteristics of its functioning in individual social contexts, as is the case, for example, when it is primarily perceived as embodied in incomprehensible technical devices or social arrangements, or when the path leading from acts of human cognition remains hidden and men are dealing directly with already objectified knowledge.

Another important conclusion in this context is that, because it is a human activity, scientific cognition or research is always an activity of a concrete cognizing subject, and, moreover, a spiritual activity that calls for intense concentration of the totality of man's creative forces. New knowledge is always a leap, a departure beyond the boundaries of what exists and has been cognized, and such a leap is beyond the capacity of exclusively the cognitive abilities of man. Only individuals who apply these abilities in cognition are capable of obtaining new knowledge.

This requires some clarification. Kant, as is known, differentiated sharply between theoretical and practical reason. He sought to identify sphere within which theoretical reason is autonomous and at the same time to show that theoretical reason, which deals only with the phenomenon, rather than the "thing in itself", cannot move beyond that sphere. Subsequently, neo-Kantians and neo-positivists, having rejected the "thing in itself", stressed and absolutized that limitation of the sphere of theoretical reason, i.e., that of scientific thinking. As a result, it was not man—possessing, in addition to cognitive abilities, also interests, strivings, and will—but a certain abstraction, a "pure" subject existing only within the framework of a similarly "pure" cognitive, that is theoretical, relation to the object, that came to be viewed as the subject of cognition.[1] Marx, however, considered that "men do not at all begin with 'standing in this theoretical realtionship *to the things of the external world*'. They begin . . . not by 'standing' in a certain relationship, but by *behaving actively*".[2]

Idealistic interpretations of cognition rest on the fact that, as an activity requiring highly specialized training and governed by specific norms and guidelines, scientific work presupposes applications of human powers and abilities that are organized and structured in specific ways, rather than arbitrarily. Let us also note that the style of presenting the results of cognitive activities in textbooks, monographs, and specialized scientific journals usually requires such a desiccation of findings that the view expressed by some cyberneticians that computers will eventually be able to engage in cognitive activities may seem not so unrealistic after all. . .

However, a view of scientific cognition as merely the application, in Marx's words, of a force of Nature tamed in a special way[3] offers very limited possibilities for understanding the

[1] In that connection see V. A. Lektorsky, *Subject. Object. Cognition,* Progress Publishers, Moscow, 1984.

[2] Karl Marx, "Randglossen zu Adolph Wagners 'Lehrbuch der politischen Ökonomie' ", Marx/Engels, *Werke,* Bd. 19, Dietz Verlag, Berlin, 1962, p. 362.

[3] Karl Marx, *Grundrisse. . .,* p. 505.

full depth, multi-layered and dialectically contradictory character of cognitive activity. Only if we follow the logical shortcuts in the chain of the already objectified results of such activities could we be satisfied with the abstraction of a "cognizing subject" as someone not of flesh and blood yet possessing those partial functions and characteristics that are needed for an equally abstract "cognizing relation". And it is in the latter case that it could be asserted that scientific cognition is fully isolated from values and from the entire content of human existence.

Viewing science, like any other sphere of culture or any other sphere of human activity, as a "functional subsystem" interacting with other subsystems in a purely external manner, we unavoidably also approach scientific cognition as an activity of a partial subject. Like any creative activity, however, scientific cognition inevitably demands from man his complete essential powers; it is also one of the ways which human culture has created for further developing and asserting these powers. This is why each concrete result of cognitive activity is not an impassive recording through cognitive abilities of individual aspects of the surrounding world, but always a human achievement, a solution to a problem that has been addressed and made meaningful by man himself. Unlike a computer solving problems defined by programmers, a cognizing subject can solve a problem, even one externally defined, only when he perceives it as his own. But if that is so, that is, if cognitive activity is goal-oriented and purposeful, then it unavoidably acquires a moral and ethical content, for awareness of an objective also involves its justification. In what way this justification is arrived at in each specific case—whether by turning to the logic of science's development, the needs of society, the demands of the employer, or in some other way—is, in the present context, irrelevant. The main point is that the act of cognition is thus immersed in a value-charged rather than value-neutral atmosphere.

The human dimension of scientific cognition is expressed not only in the fact that cognition is carried out by man, but also in that it is carried out *for* man. What I have in mind is not the applied functions of science—important as they are, they have already been discussed in the preceding section—but rather the point that the knowledge which has been acquired by man must be such that it could also be perceived and assimilated by other persons. Marx wrote: "But also when I am active *scientifically*, etc.—an activity which I can seldom perform in direct community with others—then my activity is *social*, because I perform it as a *man*."[1]

The proposition that scientific activity is social in nature also implies that scientific knowledge is commensurate with man and that it must be in principle accessible to human perception, understanding and comprehension. At the same time, scientific knowledge represents a substratum within which and in connection with which human communication takes place. Of course, as a rule that social communication is specialized and calls for subjects that are appropriately trained, but in any case these are human subjects.[2] In this regard the role of scientific knowledge is a dual one: on the one hand, as it develops and differentiates, science produces an increasing number of narrowly specialized scientific communities, among which communication is extremely difficult. At the same time, as scientific knowledge becomes part of culture, and not only of professional and national cultures but of that of the entire human race, science becomes one of the factors that open possibilities for mutual understanding. The latter element is made especially important by the fact that the revolution in science and technology continuously expands the aggregate of scientific knowledge vital in modern man's activities in a wide variety of spheres, the knowledge that he receives through the system of higher and general education.

Thus the knowledge concerning the world that is obtained by a subject is oriented on being acquired by other subjects, regardless of the extent to which the subject is aware of such an orientation. Apparently that inter-subject character of scientific knowledge is associated with the fact that scientific cognition is governed by methodological norms that each subject assimilates in the course of his own professional training, rather than reinvents anew. But since this knowledge is addressed to others it becomes possible to establish the inherent aspects of

---

[1] Karl Marx, "Economic and Philosophic Manuscripts of 1844", *op. cit.*, p. 298.

[2] Some of the characteristics of scientific communication are considered in *Kommunikatsiya v sovremennoi nauke* (Communication in Modern Science), Progress Publishers, Moscow, 1976.

social *interaction* and social activity within cognitive activities. Accordingly, from this angle as well we return to the interrelationship between scientific cognition and values, that is, the socio-ethical characteristics of scientific activity.

It should also be noted that science, and especially modern science, operates not only as cognition by man and for the benefit of man. Increasingly, modern science has also been making man an *object* of cognition. Without considering all the causes of that phenomenon or all its consequences in the development of science and philosophy, we would only stress in the present context that it is precisely in that sector that the fundamental problems of modern science are concentrated, and that what their solution requires most is a socio-ethical, humanistic analysis. It is thus obvious that it is impossible, and even dangerous for science as well as for man, to disconnect the linkages between scientific cognition and human values. To clarify, let us return to an earlier point.

We have seen that science is becoming a decisive factor promoting social transformations as well as better education and cultural development of society and the individual. In that sense, it is a force contributing to the development of man's creative potential and talents. However, as Marx had noted, in a society of class antagonisms even the pure light of science cannot shine otherwise than against the background of ignorance. Thus, modern science often plays only a minor role in the lives of most individuals, even in industrially developed capitalist countries. Approximately fifty million persons die from hunger in the world each year; 900 million are illiterate. While contributing to the growth of knowledge, science produces a still greater alienation of man, something that has already assumed threatening forms under capitalism, while mass-scale "scientific production" brings to life the same type of "partial worker" as does large-scale industrial production. Moreover, as the achievements of modern science are applied to traditional technologies, this intensifies the whole range of today's global problems, particularly that of society's interaction with Nature. Most importantly, when placed in the service of militarism science contributes to a suicidal arms race that is leading the world to a thermonuclear war.

These factors, which accompany the development of science

in antagonistic societies, explain what is not simply a disillusionment with science but a fear of science viewed as some kind of threat. The resulting situation is often interpreted as a "crisis of science" bearing on its social and world-view foundations and guiding principles. In this context we see at least two sharply defined phenomena or tendencies that are gaining ground in the social views and guiding principles of science and that point to ways to overcome that crisis.

First, as a special form of human activity science seeks to reunite itself with that activity's immediate subject, namely, man, as reflected in the fact that from being socially alienated it is increasingly acquiring a "human dimension", that is, commensurate with human qualities and needs that are expressed not only in the form of ultimate social objectives and results, but also directly. This is due to an increasing emphasis on the role of man as a subject of scientific activity. While the power of technical means of cognition is increasing, they are nevertheless beginning to lose the "demonic" qualities that were attributed to them at early stages of the scientific and technological revolution. Science is exceedingly penetrating not only the social life of man but also his personal life, significantly transforming it and subordinating it to previously unknown and unfamiliar standards and structures. An important role in that process is also played by the fact that man, as an integral unity of social and biological qualities, is becoming one of the basic objects of scientific cognition—of both the natural sciences and the social sciences. The need for taking into account the characteristics specific to man and for conforming to certain conditions in research, expressed through a number of rules and prescriptions, substantially distorts such instruments of cognition as experiments. The process is influenced by numerous considerations that do not originate directly in the research project itself but acively influence it as something that is objectively necessary. While this makes science more dependent on society, i.e., more sociologized, that dependence could also be described as individualized or personalized, thus losing its anonymous-abstract forms.

Secondly, the tendency towards more explicit and immediate sociologization and humanization of modern science is a reflection of general processes associated with the need for socio-ethical and humanistic regulation of science and for its management on a national, regional, and, very importantly, global scale. Naturally, these tendencies are perceived of and expressed differently under capitalism as vs. socialist societies, and this, perhaps, is what is most characteristic of the development of modern science. I am referring not only to the institutional regulation and management but also to the operation of socio-ethical and humanistic principles that play a regulating role in science at the level of consciousness and self-awareness of scientists and of their morality. In philosophical terms, this may be seen as an orientation of science and of scientists based on the latter's scientific view of the world and values, and it is there that one observes something new and occasionally diametrically opposite to past tendencies in science and society.

The new developments are usually attributed primarily to a number of internal changes (in the sphere of methodology, in particular, this is expressed in a rejection of neo-positivistic precepts and approaches that reduce possibilities for scientific cognition) and also to a new understanding of the place and role of science in modern culture. This second tendency is most clearly evident today in criticism of what is known as scientism.

Some time ago C. P. Snow described two "alternative cultures"—the scientific-technological one and the artistic and humanistic culture. Subsequent discussions have shown both the unsoundness of the excessive claims of scientism, including technicism as its extreme form, and the need for a more organic interaction between science and art and the development of culture as a whole. We find, moreover, that scientism and technicism now appear to coexist with an interest in art or politics as a "hobby", while haughtily rejecting humanitarian science and culture in a larger sense. This anti-humanistic role of scientism and technicism continues to gain in prominence as the social prestige of science and technology increases.

As never before, the complex and ambivalent interrelationships prevailing in human society are now also leaving their imprint on many problems in natural sciences. These problems acquire not only new parameters and a new scope but also a new quality, as has been made evident, for example, by global problems, whose development has clearly shown the insufficiency of any scientific and technological solutions if taken in isolation from a wider world-view context or from their social and humanistic aspects and dimensions. This has been stressed from the very first by Marxist thinkers who have relied on relevant classical traditions in turning, especially in the 1960s and the 1970s, to an intensive development of integrated approaches. Non-Marxist theoreticians, on the other hand, either continue to ignore them or approach them in highly contradictory ways through methods of "trial and error", as has been the case in studies of global problems in the most recent reports to the Club of Rome.

Today, aside from studies of epistemological and logical-methodological problems of science, the analysis of its world-view-related and socio-philosophical foundations, of the social consequences of the scientific and technological revolution, and of the latter's influence on man and on the social and natural environment within which he lives today and hopefully will continue to live and develop tomorrow, acquires increasing importance. The principal phenomenon that one must have in mind is the integration processes occurring in modern science and the intensification of interaction between the social and natural sciences. Thus, an integrated and systemic (and hence dialectical) approach to the problems under study is required, as is their closer linkage with the practical tasks of social development and with solutions to the global problems of today.

Such a broad interpretation of the philosophical and social relations of science with man and society has been thoroughly substantiated and has become a tradition in Marxism. In creatively developing this tradition today, Marxists identify and study new problems possessing an integrated character and therefore calling for an integrated approach. Integrated approaches that take into consideration interactions between the social, natural and technical sciences and a close interrelationship of theory with practice and with man are now increasingly becoming a major distinctive feature in the development of science.

Scientific research is increasingly focused on "interfaces" between several disciplines. This is why there is an increasing need not only to identify complex problems calling for research, but also to design wide-ranging and ramified programmes that require a concentration of substantial numbers of scientists and significant material resources, relevance of research for practical activities and man, and an integral unity of science with modern culture as a whole.

That dialectical idea has also been asserting itself spontaneously in those philosophical-methodological interpretations of science that are not associated with Marxism and are sometimes hostile to it. The very logic of cognition and life itself impose this conclusion on scientists.

For example, in an article published in 1980 in the journal *Science* and entitled "Affinities Between Scientists and Humanists",[1] James Banner refers to a report of a Commision on the Humanities sponsored by the Rockefeller Foundation on the current state and role of the humanities in American life.[2] The report notes that humanistic knowledge has the same responsibility *vis-à-vis* natural science as the latter has *vis-à-vis* the humanities. Today, when scientists and technologists focus on issues associated with the unprecedented advances of science in transforming the environment, and when it is no longer true that questions associated with the freedom of and responsibility for scientific activities lie beyond the horizon of scientists, a dialogue with the humanists has become particularly important. It is no longer sufficient to be an expert in one's field in order to be a good scientist. Scholars working in different specialized fields have to be aware of the deep kinship and similarity of humanistic and scientific forms of knowledge. Both humanistic and scientific knowledge offer examples of the highest achievements of the human mind and spirit. Both humanistic and scientific knowledge (each in its own way) create and discover real properties of Nature and of civilization. The conclusion follows that scholars working in various disciplines must continue their cooperation in the name of cognition in general. Otherwise many things will be affected, including American culture.

I believe that if tendencies towards scientism and technicism, that unfortunately one occasionally sees under socialism as well, also develop then socialist culture will also be affected. This is why it is so important today not simply to proclaim the idea of

---

[1] See: James M. Banner, Jr., "Affinities Between Scientists and Humanists", in: *Science,* Vol. 211, No. 4477, 2 January 1981.

[2] See: *The Humanities in American Life,* University of California Press, Berkeley, 1980.

an integral unity and interaction of scientific disciplines and of overall humanistic culture but also to apply it in practice.

This is the objective of the Scientific Council on the Philosophic and Social Problems of Science and Technology of the Presidium of the USSR Academy of Sciences established in 1980. Its principal tasks include: coordinating the activities of research institutions in elaborating the philosophic and social problems of science and technology; further development and implementation of Lenin's idea of an alliance between philosophy and natural science; providing an integrated, systemic approach to this research in its close linkage with the objectives of social development of the USSR and of other countries of the socialist community as well as with a scientific perspective on global issues; and philosophical analysis of the integral unity and interrelationship of natural, social and technical sciences.

The Council's diversified coordinating activities include:

—research in the role of the world view and methodology of Marxist-Leninist philosophy in the development of science and technology at the present stage; development and practical application of the idea concerning the alliance between philosophy and specialized sciences; critical analysis of bourgeois conceptions regarding the development of science and technology today;

—research in the role of science as a social institution in the context of the revolution in science and technology, of the place and role of science in contemporary society and in cultural systems; studies of methodological problems in the administration of science, of problems of the social responsibility of scientists, of ethical factors in research activities, and of socio-psychological problems of scientific and technical creative work;

—studies of the interrelationships of man, science, and technology; of the influence of the revolution of science and technology on man's development; the analysis of the philosophical-psychological aspects of man's role in the process of labour, and of human needs; studies of the problem of the harmonious development of the personality and of the humanistic aspects of the revolution in science and technology; the analysis of philosophical and social aspects of man's interrelationships with his environment;

—the elaboration from the position of materialist dialectics of a natural-scientific approach to the study of the human race's global problems, of their social aspects, of the role of science and technology in solving them, and of principles of global modeling;

—studies of the human race's future, of the correlationships among tendencies in the revolution in science and technology and development prospects of opposite social systems—socialism and capitalism; the development of methodological foundations and principles of modeling, and of exploratory and normative forecasts of scientific and technological progress; criticism of bourgeois futurological conceptions.

Thus this integrated programme of scientific research also emphasizes the catalytic and stimulating role of philosophy. True, many problems are still insufficiently clear and require scientific discussion. I believe that the important and urgent philosophical questions that emerge in this context include the criteria of scientific effectiveness generally and its evaluation from a humanistic point of view in particular. The ongoing process of transformation of science into a direct social productive force possesses its own specific characteristics, not all of which, unfortunately, are positive. Indeed, in turning to the criterion of practical effectiveness of science we run the risk of ending up with a dangerously one-sided approach inasmuch as we would be increasingly isolating science from culture as a whole. As a result the criteria of humanistic and cultural relevance of science would be relegated to the background; we see manifestations of this, as already noted, in the divergence of the two cultures, namely, the scientific-technological one and the humanitarian culture—a phenomenon that has become particularly evident in the present context. It is only natural that forms of protest and forms of asserting some new ideals in human culture involve a convergence of the scientific-technological culture and humanitarian culture in the broader sense.

It is only proper, therefore, that we are stressing the growing effectiveness of science, having in mind its effectiveness in production and its transformation into a direct productive force. It is certainly proper that the development of science is often included in those sections of plans that refer to the development of production and of the economy. But it is unfortunate that as a result science may be to a substantial degree removed from social development as a cultural factor or that its relations with all other spheres of human culture may be ruptured. For this weakens its humane mission and humanistic significance as a definite spiritual value.

This is why many scholars believe that the interaction and integral unity of science with other spheres of life, and the integral unity of humanistic culture as a whole with the necessary development of what is referred to as humanitarian culture, which at first appears as a sphere that is less effective in practical terms, are fundamental. Clearly, a major threat to the future development of human culture will develop if we will be guided in a one-sided manner only by the development of "practically effective science". In exaggerating the practical "significance" of science we may sever the vital nerves that move the development of science as a whole and the activities of scientists in particular. Science may be transformed into a soulless Moloch maiming scholars who would then themselves become "soulless scientists". This is why humanistic, value-related aspects of the development of science are so very important today in scientists' work. These constitute an integral element in today's world of science, and we will now turn to them.

Let us leave aside the analysis of various (and occasionally contradictory) philosophical definitions of the concept of value and the origins and development of bourgeois axiology. An analysis of these already exists in Marxist literature. Let us use as the point of departure a definition that is more or less widely accepted among Marxists, namely, that a value is a specific expression of the relation between a subject and an object in which the latter's properties are subjected to an evaluation in terms of the way in which they meet the subject's needs.[1] In this context the needs of a social subject are generated by society. Accordingly, an evaluation of particular material or ideal phenomena appears as socially relevant, or as something offering man guidance in the surrounding world and enabling him to transform the latter through his creative activity. Since man's activities are

[1] Kant noted that "what everyone calls pleasant is that which occasions pleasure to him, and wonderful—that which he only likes; he calls good that which he values and approves, that is, in which he perceives an objective value." (Immanuel Kant, *Kritik der Urteilskraft*, Verlag von Felix Meiner, Leipzig, 1913, p. 47.

goal-oriented, they also possess certain axiological properties, inasmuch as their outcome is expected to correspond to a predetermined purpose. The evaluation of a specific phenomenon-in-the-making through purpose (a relation of purposefulness) shows best of all the integral unity—despite many other differences—of value-oriented approaches with scientific ones, while the goal-oriented approach is an effective method of cognition.

Marx had demonstrated this brilliantly by addressing primarily the characteristics of man's practical labour activity in the course of which a predetermined objective defines, like a law, the mode and nature of his actions. "In creating a *world of objects* by his practical activity," he wrote, "in his *work upon* inorganic nature, man proves himself a conscious species-being ... man produces even when he is free from physical need and only truly produces in freedom therefrom ... man knows how to produce in accordance with the standard of every species, and knows how to apply everywhere the inherent standard to the object. Man therefore also forms objects in accordance with the laws of beauty."[1]

Marxism extends that approach to scientific activities as well. In Marxist axiology, accordingly, values, by no means separated from scientific cognition, are viewed instead in terms of their inseparable links and interaction with it. The objective genuineness of scientific cognition is combined organically with the presence of value relations both within scientific cognition itself, which represents the achievement with the help of corresponding means of a specific objective, truth, and on the part of society, which associates science with its purposes and ideals.

In viewing scientific cognition and science as a whole from such an angle we include it into object-subject relations, and above all into the sphere of methodological analysis that establishes a correspondence between knowledge and reality.

Let us recall certain characteristics of truth. Lenin defined objective truth as such a content of human conceptions "that does not depend either on a human being or on humanity".[2] But still, what is truth? That question arises not only in its evangel-

---

[1] Karl Marx, "Economic and Philosophic Manuscripts of 1844", pp. 276, 277.

[2] V. I. Lenin, "Materialism and Empirio-Criticism", *Collected Works,* Vol. 14, 1977, p. 122.

ical interpretation—which has a certain basis—but also in a sense that we associate with an axiological relation. Let us recall Hegel, who spoke so poetically about "daring to seek truth" and about faith in the power of reason as the first condition of scientific (in his case, philosophical) pursuits. Hegel stressed that "man must respect himself and recognize himself as worthy of what is highest... The hidden essence of the Universe does not possess a power that could resist his daring in knowledge: it must open itself before him and display its riches before his eyes and allow him to enjoy them."[1] Knowledge and values thus interact organically since both are expressed through man's creative, i.e., active, essence.[2]

In science, value judgements bearing on the logical structure of knowledge, and on methodological norms and relations within scientific communities, are inherent in at least one of these spheres not only in the social sciences and humanities, where they are simply unavoidable, but also in the natural sciences, which they penetrate through their philosophical orientation and methodology. And in that sense it may be said that no science is free of values. On the other hand, the fact that cognitive activities are axiologically oriented does not deprive knowledge of the attribute of objectivity. This does not mean, of course, that in each individual case value-oriented approaches operate solely in terms of their constructive function. We know that its destructive role in science continue to generate all kinds of pseudo-scientific and unscientific constructions. This led Marx to state that "when a man seeks to *accommodate* science to a viewpoint which is derived not from science itself (however erroneous it may be) but *from outside, from alien, external interests,* then I call him *'base'*."[3]

As we see, Marx's *evaluation* of behaviour is based on the manner in which methodological norms are followed (or violated)

---

[1] Georg Wilhelm Friedrich Hegel, "Hegels Anrede an seine Zuhörer bei Eröffnung seiner Vorlesungen in Berlin, 22 October 1818", in: *Encyclopädie der Philosophischen Wissenschaften im Grundrisse,* Verlag von Felix Meiner, Leipzig, 1920, p. LXXVI.

[2] In that connection let us recall Dostoyevsky's possibly "non-rigorous" but very interesting definition: "Nothing is more poetic than truth, especially in its purest state." (F. M. Dostoyevsky, *Complete Collected Works,* Vol. II, Moscow, Nauka, p. 123. /in Russian/.)

[3] Karl Marx, *Theories of Surplus-Value,* Part II, 1975, p. 119.

in a scientist's cognitive activities. These norms have the form of specific values of which the main one is a striving for truth that blocks out interests that are not only external in relation to science but are alien to it. It is particularly important to emphasize this because when the value-related orientation of scientific cognition is motivated objectively, that is when it is itself a result of scientific knowledge, it no longer operates as something alien and hence external in relation to knowledge and is therefore unable to distort truth. This, however, is a major question that calls for separate consideration.

With regard to the aspect under discussion, it is important to note that the criterion of objective truth also extends to and bears on the scientist's philosophical and ideological orientation, his understanding of general objectives and of corresponding means of scientific cognition. In that respect as well, of course, one must object not only to scientistic assertions to the effect that truly objective knowledge is opposed to value-oriented approaches including ideological, humanistic and socio-ethical valuations (Monod), but also to post-positivistic conceptions of the "philosophy of science" (e.g., Thomas Kuhn, Paul Feyerabend) in which the problem of objective truth in science is viewed as almost altogether meaningless, and where an understanding of the truth of scientific judgement depends entirely on the adopted "paradigm".[1]

In rejecting relativistic approaches, Marxists adopt a dialectical-materialist understanding of objective truth, and of the correlation therein of the relative and the absolute. Moreover, as Lenin stressed, it is for us a certain "measure or model existing independently of mankind to which our relative knowledge approximates".[2] Characteristically, in this case Lenin uses precisely these concepts—"measure" and "model", for they quite clear-

---

[1] In particular, see: Thomas S. Kuhn, *The Structure of Scientific Revolutions,* Vol. II, No. 2, University of Chicago Press, Chicago, 1970, in which the author considers the normative role of "paradigms" (as the "constellation of group commitments", "shared examples", etc.), i.e., as values (pp. 178-180). In effect, however, he rejects the criterion of the objective truthfulness of knowledge, i.e., its correspondence to reality existing outside of us, and does not object to a characterization of these positions as relativistic (*ibid.,* pp. 198-200).

[2] V. I. Lenin, "Materialism and Empirio-Criticism", p. 137.

ly convey the normative methodological (and in that sense axiological) character of objective truth and its conception as a "*process*".[1]

It therefore appears to be possible to extend to the axiological approach certain characteristics of objective truth relating, in particular, to the dialectics of its relative and absolute aspects. In doing this we recognize not only relative values but also absolute, universal values existing as certain types of "measures" and "models". In terms of the aspect that interests us, i.e., in considering the methodological meaning of value relations, this may yield, in my opinion, criteria that distinguish true science from pseudo-scientific and anti-scientific constructions.

Science rests on a number of methodological-normative values that often possess an absolute, universal significance (e.g., general principles and laws of cognitive activity, such as those of observation, experiments, inference, etc.); thus, any kind of relativism can only dilute the criterion of the objective authenticity of knowledge and lead to subjectivism. On the other hand, the process of cognition introduces elements of relativity into truth, and thus any violation of the principle of relativity can lead to rigidity in thinking, dogmatism, and authoritarianism. It is precisely on deviations from the harmonious unity of cognition and values that pseudo-science and anti-science are based. These, in turn, always imply that the principle of objective authenticity of knowledge is also violated, which is accompanied by either a false understanding of values or a complete rejection of any values whatsoever. True, in our age this is never made explicit; in particular, the erroneousness of individual value orientations can only be determined if the latter are objectively matched with the results of cognition.

There have been many situations of this kind in the history of science. Let us recall, for example, some relatively recent pseudo-scientific and anti-scientific undertakings based on false philosophical precepts that seemed to conform to dialectical materialist world views and methodological principles. They included, in particular, attempts to reject the theory of relativity and

---

[1] See: V. I. Lenin, "Conspectus of Hegel's Book *The Science of Logic*", *Collected Works,* Vol. 38, 1961, p. 201.

cybernetics, and, in the case of genetics, even attempts to create a special science concerning heredity and mutability, a science described as "materialist-dialectical" one, as opposed to "idealistic-metaphysical" science. The above were accompanied not only by a striving to erode the boundaries between what is true and what is false within science itself, but also to switch the respective positions of the latter two components. Moreover, philosophically orienting value precepts were also subjected to the same type of distortion—thus leading directly to anti-dialectical conclusions.

A distortion of value relations in particular sciences produces an overall deformation of these values and of the value of science as such. The value of science is expressed in the general context of human activities as a means for achieving the socially significant objectives of the latter. While the above is of relevance in fundamental evaluation of science as it relates to society and to the individual, it also determines a more concrete understanding of the meaning of the socio-ethical and humanistic principles (governing norms) of scientific cognition and of their guiding role. The manner in which this applies to modern science is a separate consideration and we will not turn to that question.

"Science and humanism", "man, science and ethics", "science and morality"—we have become accustomed to combinations of these concepts in philosophical writings, political speeches, religious sermons, literary works, in monologues and dialogues on the stage, in films, television, and in public declarations of scientists. There is no longer any doubt that this reflects certain specific characteristics of modern science in its interrelationship with man and society. But what is in fact actually associated with science itself and what is brought into it due to the increased influence of culture, and of conditions and factors governing its existence and functioning in today's world?

The following important question arises: are the much-discussed socio-ethical and humanistic problems of science inherent problems of science itself—or are they problems of contemporary society, disturbed by the possibility that the consequences of the application of certain scientific discoveries, should they fall outside socio-ethical and humanistic control, would be catastrophic? Is this distinction between the objectives and aspirations of science as opposed to those of society admissible today, and what are the forms in which their genuine harmonization can be achieved? As we will see, attempts to examine already existing or possible answers to these questions in an unbiased manner point to their complexity and contradictory character rather than complete what has already been achieved and stimulate new studies.

In the preceding section, which identifies certain general prerequisites for analyzing the questions to which we have now arrived, science was discussed generally, without subdividing it, as is usually done, into so-called "pure" or basic vs. applied disciplines. In terms of those aspects that are relevant here the distinction is often defined as follows: with regard to the former, truth appears to be the highest value, while with regard to the latter it merely serves as a means for achieving some practical objective. In other words, in the basic sciences truth is a self-sufficing value, while in applied sciences it is instrumental. That is a significant distinction which may be even carried to the point where the existence of two separate, qualitatively different systems of value relations—social criteria, world-view orientations, and ethical or humanistic principles—would be assumed.

With the above as a premise, it would be easy to explain, for instance, the origins of scientistic-technicistic and pragmatic-utilitarian views (with the obvious reservation that the above are mainly characteristic of capitalist society) as brought to life, among other things, by applied sciences (as if applied sciences did not develop under socialism). But it seems to me that this would be a departure from analytical principles, and in many ways would simply ignore the problem. The problem, however, emerges in terms of all its dimensions as an urgent contemporary issue only when we cease to absolutize the distinctions between basic and applied sciences as they relate to truth and values and view them as a single whole.

It is only as a whole, and as a specific social institution, that science is of value as a means for achieving the human race's practical objectives, and this refers to both applied and basic sciences (in accordance with the principle that "there is nothing more practical than a good theory"). It is science as a whole that can carry out this fundamental social function, but only when directly oriented on seeking truth and when it recognizes the objective character of knowledge as a highest value than pursues some "optimum" of what is attainable under particular conditions, as is said to be largely true of applied sciences, including political science. The emphases in basic and applied sciences obviously differ, but it is nevertheless the common elements that unite them that matter most and generally characterize science as a search for truth and a means for guiding human activity.

In such a context an important regulating role is assumed today by the socio-ethical and humanistic principles of scientific cognition. Science in the service of man—such is, in my opinion, the only socio-ethical, humanistic orientation that offers a universal basis for evaluating science in terms of its ability to serve man and to subordinate internal objectives to the general objective of mankind's social development.

It follows that what is involved is not only the value of scientific truth but also its *cost,* with regard to which the "point of reference" is man and the benefit of man.[1] The latter, it is true, has never been defined precisely and has always been relative, with its specific historical content depending on predominant social classes and other conditions. Nevertheless, it is precisely the benefit of man that has been a certain universal value; it continues to be so interpreted today when we define the long-term interests and objectives of communism as those common to the entire human race. This is why any relativism cannot be admitted, for it would undermine the very basis of humanistic ethics. The choice cannot be between what is ethical (humane) and relatively ethical (expedient, necessary). The latter, as Albert Schweitzer rightly noted, must always be clearly perceived as at least a partial violation of some fundamental principles of ethics and humanism.[2]

---

[1] Let us recall Ivan Karamazov's often cited words in Dostoyevsky's *The Karamazov Brothers*: "And if the children's sufferings go to swell the sum of the sufferings needed in payment for truth, then I assert in advance that the whole truth is not worth the price." (F. Dostoyevsky, *The Karamazov Brothers,* Progress Publishers, Moscow, 1981, p. 374).

[2] Albert Schweitzer, *Kultur und Ethik. Kulturphilosophie,* Zweiter Teil, C. H. Beck'sche Verlagsbuchhandlung, München, 1926, p. 257.

These considerations, which at first may appear somewhat general, possess an entirely concrete content that is reflected, in particular, in the situation in contemporary science as it relates to man and society. Today the question of the eventual cost to the human race of individual discoveries in nuclear physics, molecular biology, etc., i.e, in basic rather than applied sciences, assumes particular urgency. It is easy to shift the full burden of moral responsibility to science and technology, and for them is still easier to shift it in turn to society whose needs, it is said, they meet, being guided exclusively by what is necessary, expedient, and "optimal" under the circumstances. But no one can escape the problem of moral choice; thus, the evaluation of various needed decisions—whenever they fail to meet ethical, or humanistic, norms even to the smallest extent—as a violation of these norms (and hence as an evil, even though possibly unavoidable) allow us to hold back the development of negative processes at a strictly defined level and to actively oppose them, guided by a clear perspective. It is in this respect that the position advocated here differs not only from ethical relativism and nihilism (scientism) but also from the "critique of science", based on Rousseauistic positions, which advocates that scientific and technological progress be held back. For us it is clear that it is only further comprehensive scientific and technological progress for the benefit of man that can remove the negative effects of science and of its applications. But this may be achieved only in social conditions which have the benefit of man as the highest goal.

Today these questions are being discussed in science and in philosophy primarily in connection with the need to solve concrete problems concerning the admissibility or inadmissibility of individual research activities that either directly or in their technical applications may endanger man and the human race. This refers not only to research in nuclear physics but also, to a substantial degree, molecular biology, genetics, medicine, psychology, etc., that is, everything that is directly or indirectly related to the cognition of man. These problems will be considered in the second chapter which is concerned with an analysis of the socio-ethical and humanistic principles (regulators) that operate in concrete spheres of the cognition of man.

In summarizing what has been stated concerning the interrelationship of man, science and humanism, and rephrasing Protagoras' well-known statement, it may be said that it is man who is the "measure of all sciences" and that only this humanistic orientation can create a firm basis for man's future and the future of all human culture. And from such a vantage point both the absolutization of science's significance in the life of man and that of the human race as well as for the development of a future humanistic culture, and attempts to minimize its importance, "dethrone" it, and present it as an anti-humane force, are equally dangerous. The true meaning of science emerges only in connection with other forms of human activity that form the material and spiritual culture of the human race.

In that respect a great role is played by man's labour, art, and morality. In each of these one finds a distinct manifestation of the creative essential forces of man and of the human race. The moral and humanistic foundations of science and of all human culture are acquiring a growing importance, for the apparent alternative to them is darkness, a lack of culture and a spiritual and physical collapse of the human race. There is an increasingly wide and profound awareness of this on the part of scientists, which has found expression in a number of international research projects (the UNESCO project on science and human needs, etc.). The humanistic aspects of the development of science and technology have been reflected in documents and reports of UNESCO's second World Conference on policy in the field of culture (Mexico, July-August 1982). In particular, André Danzin and Ilya Prigogine rightly stress that there should be a closer involvement between science and human values. "The message of science will then cease to be purely technical and it will be possible to understand man in all his complexity, in his need for social stability just as much as in his desire for personal freedom."[1]

In that connection traditional moral-humanistic and ethical problems of science are reemerging and becoming more urgent. There is a growing awareness that if the social responsibility of scientists and the role of moral and ethical principles in science

[1] André Danzin and Ilya Prigogine, "Are We Getting the Science We Need? A Hard Look at Science as a Trigger of Change", in: *The UNESCO Courier,* February 1982, Paris, pp. 7-8.

do not increase in a geometrical progression, then science itself will not be able to develop even at an arithmetic rate. Of course, in referring to the integral unity of science and humanism, as well as to that of the human objectives of cognition and mankind's ethical values, we do so primarily as a *principle* and a *prospect* of genuine science. In reality, as we have seen, science in its present forms is in many cases still quite distant from this. Still, we must base out position on the point that *the fundamental unity of science and humanism and science's development as a force serving mankind's progress represent one of the most urgent problems of modern development. There is no alternative to the above, either for science or for the human race.*

An awareness of this makes it possible to better evaluate the humanistic dimensions of scientific research and of its results and hence to overcome the ethical relativism and nihilism that threaten mankind. It is becoming increasingly clear that it is no longer sufficient to "wave the flag of Galileo" and that science cannot develop in a "social vacuum", that is, in isolation from world views and from socio-philosophical and ethical foundations. Far from being something external to science, moral and ethical problems are entwined in it. One is brought to this conclusion by the history of the development of science, its growing interrelationship with society, and its expanding social functions. One is led to the same conclusion if one examines the evolution of science's self-awareness, including the evolution that is taking place in what is referred to today as the *ethics of science*. This question deserves separate consideration, for it has become the subject not only of scientific polemics but also of widely diverging appraisals made by broad public opinion throughout the world. It is directly linked to the problem of freedom to engage in scientific research and of the socio-ethical and humanistic responsibility of scientists, which has become a global problem of relevance to the present and future of mankind.

3. Discussions on Science vs. Ethics: an Alternative or Interdependence? Freedom to Engage in Scientific Research and the Responsibility of Scientists; the Ethics of Science, Its Origins, Specific Characteristics, and Relation to Mankind's General Socio-Ethical Values

In turning to the mutual relations of science and ethics, and to the ethics of science itself, we enter into a sphere in which there are more questions than answers, and more polemics than already established positive solutions. We will also see that it is, unfortunately, impossible to answer these questions in an unambivalent manner at the current stage in the development of research in this field. But why "unfortunately", then? Does that mean that all questions require precisely such answers? Yet there exists an entire range of "eternal" problems of human existence with which philosophy and science are permanently concerned, and still we return again and again to the point of departure, while those questions reemerge with increasing urgency before new generations living in new historical conditions and, accordingly, resolving these "eternal" problems (that are actually concrete-historical) in new ways.

Indeed, how were the problems of the meaning of life, or of good and evil, for example, defined and solved in various ages? Is it really possible to apply to them the same "criteria of scientific validity", as, for example, to the laws of mechanics? Yet they are certainly no less important to man's existence.

In the words of Alexander Pushkin, human genius is a "friend of paradoxes". And the sharpest paradoxes arise when we turn to these "eternal" problems. They also include those that are today described as the socio-ethical and humanistic problems of man and of scientific cognition. This is why, I believe, discussions relating to these problems should be evaluated not so much in terms of whether they provide final answers to questions or whether they bring us closer to constructing certain universal codes of behaviour of scientists, as in terms of their impact on the development of social responsibility of scientists. These discussions also involve more general problems concerning, first, what is known as the dilemma of science vs. morality, and secondly, ethical codes of scientists and their relation to the general ethical values of mankind.

Studies of the correlation between science and morality often generate the idea that science is a humane and moral force in itself, and that it therefore does not need any external ethical-humanistic guidelines. In that conception the inherent humanism of science coincides with the ethicalness of truth and of the process of its cognition regardless of the social context within which the latter takes place. Such a conception of the correlation of science and morality represents an attempt to revive the ideas and conceptions that reflect the classical image of science, and are hardly acceptable today, when science has ceased to be a closed guild-like organization, as was still true in the nineteenth and early twentieth century, becoming instead a social institution that is linked to mankind's life by a thousand threads. What is more, technology and technological applications of science have developed enormously and have become more complex while the division of research into "pure" and applied has become more marked. The "classical" view seeks to separate applied research from science more generally in order to at least preserve the outward appearance of its purity and to deflect from science accusations of complicity in technological applications that may be viewed as at least partial violations of the principles of humanism and of morality.

As noted above, such a distinction is unfounded; this is why the "classical" view of the mutual relation of science and morality, which ignores the interaction of the ethical values of cognition with mankind's general ethical values that are social in nature and that largely determine the direction and nature of scientists' cognitive activities, cannot serve as a reliable compass today. Rather the opposite is true: it can disorient scientists, and remove from them the burden of moral responsibility in difficult and complex situations involving anti-humane technological applications of science, insofar as the scholars would be able to appease their conscience through the thought that they were not involved in such applications which are, moreover, distinct from science.

In that connection I would like to note that if one turns to the history of the issue regarding the correlationship of science, morality and humanism, it cannot be said that modern science defines these problems in ways that are much more profound and diverse than previously. It seems to me that there have been wave-like movements in the history of modern science and even returns to earlier achievements as regards these questions, and also that different stimuli existed in different periods as did different "trigger mechanisms" activating the discussions.

In my view, the above requires a close analysis from the standpoint of the history of science. I would like to express the following view which may perhaps be controversial. It would appear that a turn towards these problems is not associated directly with a growing sense of social responsibility among modern scientists. It would seem that the explosion, and subsequently the decline, of interest towards socio-ethical and humanistic problems of science are due to some other causes that may be even more powerful than merely a sense of social responsibility on the part of scientists in a particular period.

There one sees the reverse side of the fact that science today has become a multi-component and multi-factor phenomenon. It is now truly a social institution in modern society; accordingly, its development depends on the characteristics of particular periods in society's history—a factor that explains, in my opinion, the rises and falls in interest in humanistic problems. If one considers distant history and compares it with modern times, what one can find is not an increase in the sense of responsibility of scientists, but rather an increased dependence of science on society, its ideology and its politics, and accordingly an increasingly close relation between science and scientists, on the one hand, and the rises and falls that society experiences, on the other. This is why in many cases what we observe is not an increasing sense of responsibility among scientists today, but rather a growing sense of guilt—something that was absent at earlier stages in science's development and even in the age of classical science in the nineteenth century.

Whether this is actually so is, of course, debatable. I would only like to remove from modern scientists a certain aura of allegedly greater moral responsibility by comparison with those who lived earlier and to say that perhaps we should take a more positive view of scientists who lived in the past. I believe that the moral and humanistic level of nineteenth-century scientists, for example, was higher on the average than today. Once again, the above is associated with the specific nature of modern science. Today the latter has become a form of mass production, while in the past scientists were recruited primarily from among persons of outstanding talent, rather than of average abilities as is the case today.

This has been reflected to a certain extent in the development of science, including its socio-ethical and humanistic aspects. In any event, when turning to literary, historical, autobiographic and other sources I often find more questions in the past that call for reflection than when I study corresponding materials pertaining to modern science. Very much depends on the way in which science is understood more generally, and on the way in which it is actually conceived at various stages of its development.

Today there are many references to various types of science, but I would like to stress that the humanistic question concerning the *purpose of science*, which to us is basic, was formulated a long time ago. It may be that today it is sometimes formulated and solved less adequately than should be the case. This does not mean that there is a degradation of thought by comparison with what existed previously in the history of science. But, as noted, at the present stage in the development of science that question is being solved less adequately than it should be. And if one turns to nineteenth-century science one could say that even though the socio-ethical and humanistic problems of science were then in many respects only incipient, they were defined more sharply in philosophy than today.

As to the present, I would repeat that what we witness is probably a much greater influence of politics and ideology than of ethics and morality; we define many things as ethical phenomena while in actual fact their nature is political and ideological, and this suppresses scientists' independent ethical decisions.

The only conclusion that follows is that independent research has to be developed and solutions elaborated to today's socio-ethical problems of science with due account of past experience and of its relevance to the modern age. This will make it possible to avoid repetitions and to move forward in formulating and solving new questions. The above applies not only to enlightening conceptions which originated with Socrates and which state that knowledge and virtue are inseparable, but also to their antipode, namely, the Rousseauistic conceptions of the incompatibility of knowledge with virtue.[1] Let us consider how this has been reflected in modern discussions concerning the correlation of science and ethics, and of ideals and norms of science and mankind's moral values. Let us try to trace the changes in the interpretations of that question which have taken place in recent decades and the different views that reflect, to a certain degree, the fact that their adherents came from different regions of the world.

A convenient point of reference for the purposes of this analysis is provided by Henry Margenau, a prominent American physicist and philosopher, in an article entitled "Western Culture, Scientific Method and the Problem of Ethics", first published in 1947.[2] In a general characterization of Western culture Margenau presents a point of view, admittedly relatively unpopular, which he nevertheless seeks to substantiate. This is "a view that places scientific method in the center of interest, leaving philosophy, aesthetics and religion in the background of the cultural scene."[3] Thus, he writes, the basic features of Western culture are reflected most graphically in the methodology of science. Science, meanwhile, is interpreted quite narrowly as "rational, or theoretical, science" so that it excludes social science and psychology, whose "present state is more like the condition of geometry in pre-Pythagorean times"[4] as well as history and anthropology which "have hitherto been disciplines largely marked by creative impotence, except in so far as they give satisfaction to scholars".[5]

It is not surprising that within the framework of this scientistic point of view the question of the correlation of science and ethics (let us note that Margenau stresses his firm conviction that ethics and science are compatible) is considered primarily from

[1] This idea was formulated by Jean Jacques Rousseau in his essay "Si le progrès des sciences et des arts a contribué à corrompre ou à épurer les moeurs?".

[2] Henry Margenau, "Western Culture, Scientific Method and the Problem of Ethics", *Physics and Philosophy: Selected Essays,* D. Reidel Publishing Co., Dordrecht (Holland), Boston (USA), London (England), 1978, pp. 225-240.

[3] *Ibid.,* p. 225.

[4] *Ibid.,* p. 228.

[5] *Ibid.,* p. 226.

the point of view of whether "the method of science, in particular of theoretical science, [can] be applied in ethics".[1] While Margenau considers that science is ethically neutral, for it may only operate as a means after an ethical choice has been made, he nevertheless offers an affirmative answer to the question of whether the method of science can be applied to ethics, thus essentially proposing to create an ethics on a scientific basis. Science thus serves as a model for reforming ethics, while the latter in its current state is viewed as backward and as failing to meet the standards of "theoretical science". Science, on the other hand, is viewed by Margenau as the determining and supreme value in European culture.

Such views were widely held in the 1940s and 1950s, but already in the 1960s the situation began to change, as became evident in Margenau's own views. In a study published in 1966, while not abandoning the proposition that science is morally neutral, he uses substantially different reasoning when he speaks of the predicament faced by modern man: "Enabled by science to accomplish ever greater and more massive tasks, saddled with an increasingly awesome responsibility, he discovers that his very source of power leaves him without moral guidance. This is why the scientist today, more than ever, needs the council and the guidance of the humanist."[2] As we see, the idea is not that ethics be reconstructed in accordance with the norms of "theoretical science" but rather that science itself needs humanistic regulating principles originating in the humanitarian knowledge that had been neglected previously.

In the same context, the evolution of Victor Weisskopf's views that may be seen in two studies concerned with the ethical problems of science is of some interest. The time interval separating the works is quite modest: "Science and Ethics" was written in 1968, while "The Significance of Science" was written in 1970-71. While in the first of these articles one finds traces of the academic tone of preceding decades, the second is written quite emotionally as befits a discussion of a problem arousing immediate interest. One of the second article's epigraphs is taken from the Bible's Book of Ecclesiastes: "For in much wisdom *is* much grief: and he that increaseth

[1] *Ibid.*, p. 235.

[2] Henry Margenau, "The Pursuit of Significance", *op. cit.*, p. 353.

knowledge increaseth sorrow." (Eccles. I : 18). Weisskopf notes that very often the methods and findings of the natural sciences are insufficient; in other cases they are altogether inapplicable to the resolution of basic human problems. And while "new scientific knowledge is neither good nor bad",[1] there is nevertheless "a broad spectrum of relations, philosophic, social, and ethical, by which science influences and is influenced by society. The significance of science becomes evident in the numerous, often contradictory, ways in which it interacts with the affairs of man."[2]

Obviously, such a position of science within society makes it far from neutral in an ethical sense, and that being so, a considerable share of responsibility for the world's future, falls on scientists. Naturally this does not relieve them of responsibility for developing scientific knowledge as a cultural heritage, an eternal treasure of the human race and an important social asset, just as it increases, rather than annuls, their responsibility for disseminating knowledge through teaching and popularization.

This, then, is Weisskopf's point of view. Attractive though it may be, and while it raises some genuinely urgent issues, such as the struggle of scientists against the arms race and the need for their active participation in harmonizing man's relations with Nature, it should nevertheless be noted that Weisskopf is unable to identify the social forces and social conditions that could offer a real solution.

Let us now turn to conceptions involving most intense discussions of problems raised by the development of biology and of scientific disciplines concerned with human behaviour. These problems have lent a new intensity as well as a somewhat different approach to the issue of the correlation of science and ethics. The attention of the academic community has been drawn to the views of Jacques Monod, a prominent molecular biologist, who has unfortunately turned against dialectical materialism, completely distorting it in his notorious book *Le Hasard et la Nècessité*. Monod's overall philosophic conception is sufficiently well-known and has been analyzed from various positions, includ-

---

[1] Victor F. Weisskopf, *Physics in the Twentieth Century: Selected Essays*, the MIT Press, Cambridge, 1972, p. 343.

[2] *Ibid.*, p. 362.

ing that of Marxism.[1] Nevertheless, it is appropriate to turn briefly to his approach to the correlation of science and ethics, since Western authors continue to repeat his views in scientistic-technocratic conceptions.

Monod believed that man possesses an imborn, genetically encoded need to search for the meaning of existence, a need that has brought to life all known myths, religions, philosophic systems, and even science itself. For hundreds of thousands of years man's existence was identified with that of his group or tribe, outside of which he could not exist. Because these groups or tribes were of necessity cohesive, the laws underlying that cohesion possessed exceptional subjective importance and were not questioned.

Monod claimed that all of the above had influenced the genetic evolution of the human mind's "inherent categories" and created a need for mythical explanations. The creation of myths, religions, and numerous philosophical systems is the price that man must pay in order to survive as a social animal. The role of such explanations is to "alleviate one's anguish" (satisfy a need to find meaning in existence); they also link human history to the history of the cosmos by making man's existence meaningful and establishing his role in Nature. According to Monod, over many thousands of years these explanations were embodied in an "animistic tradition".

Monod asserted that because of its "unscientific" and "ideological" character dialectical and historical materialism carries on that tradition. Breaking with that tradition and asserting objective cognition as the only source of genuine truth constituted the programme for a new stage in the "age of ideas" proclaimed by Monod. He believed that modern science breaks man's ancient "animistic ties" to Nature, that, accordingly, it cannot alleviate his "inborn anguish" and that it therefore has a role to play in man's practical activities rather than in the human soul.

A conclusive break with the "animistic tradition" presupposes, according to Monod, a complete revision of the foundations of ethics and of the entire system of values that has guided the human race until now. More, it implies a basically new correlation between ethics and knowledge, and between objective truth

[1] I. T. Frolov, *Dialectique et Etique en Biologie,* Moscow, Progress Publishers, 1978.

and a value-based approach. While in "animistic conceptions" ethics and knowledge are viewed as two aspects of the same reality, the requirements of objectivity establish a fundamental distinction between them, a distinction that is unavoidable, Monod believed, in studying truth itself; this is why knowledge precludes any value-based judgement (other than "epistemological value"). Meanwhile, because of its non-objective essence, ethics is forever banned from the field of knowledge.

At the same time, Monod asserted that even though true knowledge does not recognize values, establishing it nevertheless calls for judgement, or rather a value-based axiom, and that the very adoption of the "postulate of objectivity" as a condition of genuine cognition represents a specific value which in itself represents objective knowledge. Adopting the "postulate of objectivity", accordingly, means defining the fundamental postulate of ethics, that is, the ethics of knowledge. The fundamental distinction between the latter and "animistic" ethics is based on the premise that it is an ethical choice of the original value that forms the basis of cognition.

Monod believed that no system of values can claim to represent genuine ethics unless it proposes an ideal that is transcendental in relation to the individual to such an extent as to justify self-sacrifice if necessary. The ethics of cognition could meet that requirement for transcendence by defining it as a value. Still, it does not reject humanism either, for it recognizes in man the creator and exponent of the transcendental.

To Monod, the ethics of cognition requires that man be seen as a being who simultaneously belongs to the biosphere and to the realm of ideas, who is tormented by this dualism, and who develops by seeking to express himself in art, poetry, and love. Monod completes that hymn in praise of the ethics of cognition with the conclusion that his is the only rational and ideal position. He juxtaposed it to Marxism, allegedly based not on science but on "animistic ideology", in the belief that inherent in historical materialism was a complete confusion of the categories of value and cognition. In opposing the latter, what Monod essentially defended was a scientistic conception of "pure knowledge" that is distinct from the ideals of individuals and wholly alienated from man, his subjective world, and his needs and strivings.

A somewhat different position, albeit leading to similar conclusions, was taken by Gunther S. Stent, a prominent American geneticist who advocates the idea that there is a conflict between European science and European morality.[1] Stent's interpretation addresses the rational foundations of morality, an aspect previously studied by Kant. Morality and science possess a common basis, that of human reason. Paradoxical in nature, human reason "correlates" morality and science on the basis of their fundamental similarity, inherent disparities and mutual incompatibility. Their inherent disparities do not preclude the development of a superficially coherent image of reality in order to elaborate a rational line of behaviour in everyday life. The disparities become apparent only in an in-depth analysis by scientists and philosophers. The emerging disparities and lack of clarity can be eliminated only by changing certain fundamental intuitive postulates concerning the world. Such changes may produce some "local" coherence, but will have serious consequences for cognition for they will contribute to man's alienation from that reality which he constructs in his everyday life.

Stent sought to substantiate these ideas by turning to the numerous sources relating to science's and philosophy's past and present. A particular role in his argument is played by Niels Bohr's ideas, and especially his principle of "complementarity" on the basis of which Bohr developed quantum physics. Subsequently expanded, among others, by Bohr himself, today, in Stent's view, that concept represents a general theory of the nature of human cognition. The concept of "complementarity", encompasses three interconnected and yet substantially different aspects, namely, the instrumental, metaphysical and intuitive aspects.

The intuitive aspect is directly linked to the subject of science's relation to ethics. Referring to the works of Piaget, Stent asserts that epistemological concepts emerge automatically in each normal person's early childhood as a result of a dialectical interrelation between his developing nervous system and the physical world. Accordingly, these concepts are not a product of culture, for they are genuinely intuitive; it is by developing

[1] Gunther S. Stent, "The Dilemma of Science and Morals", in: *Genetics,* Vol. 78, September 1974, pp. 41-51. See also Stent's report to the 16th World Philosophic Congress (Düsseldorf, 1978).

these concepts that a child turns into a normal human being.

In Stent's opinion a scientific approach to the world, with the help of which we seek to understand the reality of objects governed by relations of causality, is only one of the two aspects of the global intuitive ideology on which all our experience is based. The second aspect is an ethical approach which we use to elaborate the norms of interpersonal relations. That global ideology possesses many concrete expressions, both religious and secular. In most cases, however, the distinction between scientific and ethical approaches is either not recognized or deliberately rejected. As a result, the global ideology remains relatively free from both internal conflicts and conflicts between science and ethics, a factor that makes it impossible for science to develop as an autonomous intellectual activity, and thus creates obstacles to our understanding of the world of objects and to establishing our dominance over that world. To Stent, the intuitive conceptions that underly an ethical approach involve a number of concealed presumptions which create internal discrepancies in the image of reality. A scientific approach makes it possible to identify the presumptions that are concealed in our concepts of time, causality, etc. Likewise, philosophers concerned with problems of morality had to elucidate the presumptions embodied in our concepts of morality. Stent believes that such studies confirm the Kantian philosophy of morality, according to which our ethical approach to the world reflects what is biologically given rather than what is conditioned socially or philosophically. Consequently, moral principles are inherent in reason and their construction represents a child's transformation into an adult.

Stent then raises the following question: how can one resolve the paradoxes that emerge on the basis of the intuitive "complementarity" of the human reason? Like Bohr, he proposes that we turn to Eastern philosophy, and in particular to three of its teachings, namely, Buddhism, Taoism and Confucianism, where the disturbing division between the subject and the object is absent. The world is viewed as an organic whole; each person possesses from birth a potential ability to penetrate into "things in themselves". An ability for direct cognition of reality follows from intuition rather than from logical reflection and inference. Similarly, cognition of the world is viewed not as an ability to cognize a particular structure of objective truths, but as an understanding of one's own state of mind and an evaluation of one's own internal feelings. Accordingly, wisdom implies knowledge of one's own self; together with wisdom one achieves knowledge of the Way or Tao, which allows man to strive towards continuous harmonization with an ever expanding system of relations. One cannot trust reason for it constitutes an obstacle to gaining knowledge of the Way. Objective truth cannot be found along the Way.

Thus, Stent concludes, the three Eastern teachings have produced a point of view on moral problems that differs from the Western one. What is more, he attributes the emergence and development of movements to protect the environment to the growing influence of these teachings on public opinion in the West. According to Eastern teachings, he notes, virtue alone does not imply an ability to make moral judgements that are objectively correct. Moral behaviour is not viewed in this context as something that implies a choice and a responsibility, since it is considered that the (normative) Way to social harmony does not have any crossroads. One can follow the Way, one can stop or even deviate from it, but there are no alternative directions. Accordingly, the central moral question in this regard is not that of man's responsibility for the actions that he chose to undertake under the guidance of his own free will, but rather whether man has studied the Way correctly and whether he has the desire to persist in learning. Since the knowledge of the Way is already within us, virtue derives from gaining knowledge of oneself. Individuals can perfect themselves through self-development.

Stent thus arrives at the following conclusion. Having acquired reason during childhood, man can also suppress it and in that way overcome the painful paradoxes of his situation deriving from rationality; but such a simplified freedom from conflicts, it would appear, will have been purchased at a very high price, namely, that of destroying the foundations of ethics and of science.

It should be noted that such views possess a certain currency in the West, in particular among scientists. One is left with the impression that their adherents view culture not as a developing integral entity which is dialectically contradictory, but as some kind of mechanical mixture from which one may arbitrarily take particular components and replace them by new ones borrowed from outside. In opposing them Marxists show that the dilemma of science vs. morality may be resolved through *scientific* means, without turning to intuitive mystical precepts, and that such a solution, far from being at variance with the traditions of culture, on the contrary, follows organically from these traditions, including those of European culture. Marxism incorporates these *world-wide* cultural traditions (and not only those of European or Eastern culture). Of course within the framework of Marxism there may be different approaches to a concrete understanding of the problem of science vs. ethics.

Let us turn, for example, to a discussion that I believe is typical in that respect, as it has in many ways stimulated further development of Marxist theory concerning the correlation between science and morals, and for that reason it deserves attention. I have in mind the round-table discussion of the journal *Voprosy filosofii* on the subject of "Science, Ethics, and Humanism".[1]

Participants in the discussion included philosophers and scientists who proposed various approaches and concrete interpretations of the correlation between science and ethics. Yet all of them based their position on the general world-view principles of Marxist-Leninist science. Vladimir Engelgardt, a leading biochemist, expressed, in particular, the thought that by itself science does not create ethical values; it creates only one value, knowledge, which by overcoming ignorance produces an ordering of our conceptions. Since cognition of the world is an endless process, it follows, in Engelgardt's view, that knowledge as a value created by science differs from all other values created by man's inner world precisely in that it grows incessantly. As for ethical values, in his opinion, these are, on the contrary, definitely dimensional, permanent, and inviolable.

While agreeing with Engelgardt that natural sciences do not provide grounds for developing ethical conceptions, Alexander Malinovsky stressed that ethical values play a major role in the development of science itself, contributing to its growing effectiveness. He drew attention to the point that prominent scientists

[1] See: *Voprosy filosofii,* Nos. 6, 8, 1973.

creating major scientific values are generally very kind and highly ethical persons. Teodor Oizerman, a historian of philosophy, stated that science should be viewed not only as a means of cognition, but also in terms of the capacity of scientific activity to form a specific personality type and corresponding ethical norms, to produce higher ethical values. The emergence of men for whom intellectual search is the highest good revolutionized ethics.

Bruno Pontecorvo, a physicist, did not agree. Like Engelgardt, he assumed that science cannot automatically create a more human personality type simply because the person involved is engaged in scientific work. In elaborating, Engelgardt added that scientists are not in a privileged position, by comparison with persons involved in other types of activity, with regard to the cultivation of ethical values. Mikhail Volkenshtein, a biophysicist, drew attention to the fact that science is not merely a sum total of knowledge; it also involves creativity, corresponding social institutions, and forms of interrelationship of man with both himself and the surrounding world. In that respect, in his opinion, science is inseparably linked to ethics. The basic ethical categories with which science as a form of human activity is concerned are, in his opinion, truth (and the search for truth) and harmony (and the search for harmony) in the surrounding world. Both have an immediate ethical content, for truth is an ethical category. Bonifaty Kedrov, a philosopher, supported Volkenshtein's views in this respect. He noted that, while playing a specific role in social life, science by itself does not create ethical values but uncovers objective truth. It provides principles that serve some classes for one set of purposes and different purposes for others; thus revealed is the social orientation of science, which, in the final analysis, cannot be separated from its contents.

The question was raised concerning the humanistic function of science that has to be reflected in solutions of its basic problems associated with man, and also concerning the frequent noncorrespondence of the actual image of modern science with that ideal model. Since the development of science is in many ways autonomous the question arises whether there exist internal scientific mechanisms for its humanization or whether that function is external to science. That question was clearly defined by

Nelya Motroshilova, a specialist in the science of science. She emphasized the need to establish whether or not science does have a potential for internal humanistic self-control, whether humanistic guarantees are inherent in scientific activity as such of whether, on the contrary, the situation is one in which science itself creates neither good nor evil, while specific external forces lying beyond the precepts and techniques of research proper play a decisive role in regulating the activities of scientists. In her opinion, modern developments are producing encouraging tendencies in this respect, and there are grounds to believe that considerations of humanism, too, will increasingly penetrate into the very structure of science as guiding norms influencing concrete scientific research activities.

Mikhail Lifshits, a prominent specialist in esthetics, drew attention to contradictions that may develop between a striving for truth, on the one hand, and the ethical aspect of cognition and a humanistic orientation, on the other. He stated that in the interest of achieving the full scope of truth as the objective of knowledge individual aspects of scientific research activities may well be restricted. Science—not as a striving for truth, but as a social force—requires social control. Volkenshtein agreed with that idea and asked: what is it that should be regulated, science itself or its applications? In his opinion, controls should be established with regard to applications of science. In that connection Oizerman stressed that applications of science can themselves acquire the nature of research.

Vladimir Engelgardt also noted the significance of the social responsibility of scientists. With regard to the issue of good and evil as fruits of science he noted that these become evident in an analysis of actual interrelationships between science and its technological applications. It is there that a situation develops that he humorously described in terms of exclamations of surprise or concern. Some scientists are elated by the wonder of having changed the nature of microbes, making them perform technological processes. Others express concern over the danger that work on microbes will eventually be followed by attempts to change man's psychology. Engelgardt expressed his conviction that the correct way will be found. In his opinion, the possibility of arriving at positive results is inherent in the nature of scientific research, while the dangers lie essentially only in the operation of human society and can be prevented by the latter.

In defining the process of cognition in terms of philosophy as something that is "dimensionless", Merab Mamardashvili, a philosopher, argued that science is a human value inasmuch as no specific value dimension can be attached to its contents and to the corresponding states of human consciousness. In his view only one thing matters in science, namely, to derive additional knowledge from both already available knowledge and observations.

Discussions of that issue revealed differences in views. Eric Soloviev, a philosopher, noted that a specific feature of current discussions concerning the potential of humanistically oriented science is an extremely close association between and even identification of science and technology. Thus, ethical-humanistic evaluations of scientific knowledge, even though merged into a single complex today, in fact reflect fundamentally distinct approaches to reality. Technology is primarily a "culture of means"; accordingly, mastering the reality of technology requires viewing technology in terms of its suitability for human needs, usefulness, measurability, and instrumental role. As for science, its basic guiding principle is that of cognition of reality "such as it is", such as it exists independently of human interests and needs, and even in spite of them. Science views reality as a "self-being" and in an extreme case requires preserving it and reproducing it under conditions of a technological-pragmatic expansion. The concerns voiced so urgently in modern ecology, for example, appear as visible expressions of a general inherent pathos in science, the pathos of a certain concern with preserving the world as it exists, independently of human will and consciousness. In contract, Vladislav Kelle, a specialist on historical materialism, stressed that science can only develop as an endless knowledge-producing process and that this underlies both its neutrality with regard to good and evil and its internal ethical sterility. At the same time, the level of knowledge achieved by science requires the establishment of some control norms which would be not merely those needed for producing scientific knowledge but also those that would limit specific branches of the development of science.

Since ample instances have been documented of inhumane uses, external in relation to science, of scientific achievements and of science's dependence on socio-cultural backgrounds and the mechanisms used to manage both science and the consumption of its achievements, factors external to science are forecast as measures of control. In that connection, Soloviev observed in the course of the discussion that ethically oriented forms of control are required not so much for science as for its social organization and modes of its association with technology and the economy. Forecasts reveal a new type of activity, of experiments and of organization of research that requires scientists to demonstrate that applications of science are socially relevant and are not in contradiction with vital human values.

In that connection, certain elements in programmes of democratic and radical left-wing movements in the West are of major interest. I have in mind the idea of a wide democratic control over the institutions that direct the social development of science, the nature of its applications and the general orientation of scientific research. The effectiveness of such control, which also includes ethical and humanistic criteria, depends substantially on whether and to what extent it is authoritative, on the extent to which leading scientists, motivated by their concern with problems of the social organization of science, actively participate in it, and the extent to which science's alliance with democracy will be firm.

This, then, is the spectrum of problems that were identified during the discussion of the relationship between science and ethics organized by *Voprosy filosofii.* Their consideration continued in a number of subsequent publications and discussions published in *Voprosy filosofii* itself as well as in *Istoriya yestestvoznaniya i tekhniki* (History of Natural Science and of Technology), a journal founded in 1980 (see No. 4, 1980, No. 2, 1982, other issues).

Science's socio-ethical problems are now also widely discussed at the international level. In particular, one should mention the diversified activities of UNESCO which has organized a number of studies, including those on problems of ethics and genetics.[1] As already mentioned, these problems were also discussed

---

[1] See also I. Frolov, "Science for Man (Social Philosophical and Ethical Problems of Genetic Engineering)". This is a UNESCO-commissioned report considered at conferences on *Genetics and Ethics,* held on Oct. 10-14, 1977 in Madrid, and on Feb. 28-March 3, 1978 in Amsterdam.

at the 16th World Congress of Philosophy. They were reflected, too, in discussions at the 7th International Congress of Logic, Methodology and Philosophy of Science (Salzburg, Austria, 1983). In particular, Jean Ladrière (Belgium) noted in his report entitled "The Ethical Dimension of Knowledge" that science may be viewed as an action system. That system, in turn, may be linked to other action systems, mainly to technological systems, and economic and political systems. Many of the ethical problems raised by science may actually be viewed as projections of science's action system into these other systems. In Ladrière's view, however, the genuine role of the ethicist of science involves science itself, rather than its applications in other spheres, for an ethical problem emerges only when there is responsibility. Responsibility in science has two dimensions. One of them involves its own development. The other is an indirect responsibility in a situation with many *"possibilities"*, where it makes possible certain actions outside the sphere of science, but without which the latter could not exist. In Ladrière's view, one may speak of science's internal and external responsibility. If an ethics of science does exist it cannot take a form of external determination by rules which would be imposed by principles that are external to scientific activity. Science's ethics must have the form of internally normative scientific activity.

It is Ladrière's opinion, however, that the ethics of science itself possesses an integrating requirement. This presupposes that science is inscribed in a certain context. And that, in turn, prescribes to science the task of seeing itself as a partial contribution to a structure that seeks to establish a generalized reflection of reality. It follows, Ladrière concludes, that science ultimately possesses an ethical dimension only to the extent that the latter is needed for its own self-determination in the process of such an "integration".

A different position was presented at the Salzburg Congress by Marian Przełęcki (Poland) in a contribution entitled "Ethical Aspects of Non-Ethical Theories". Przełęcki believes that the ethical aspects of scientific theories may be determined in various ways, and he considered only one of the meanings of that concept in his report. The real object of a moral evaluation is human action, and therefore it is not theories but rather acts of proposing them that may be evaluated in moral terms. In the case of scientific (non-ethical) theories, the act itself of proposing them is morally neutral. This, of course, does not imply that the scientist should be indifferent to the social uses of his theories. He must be interested, but as a citizen rather than as a scientist, Przełęcki concludes.

As we see, the question of ethical principles, whether internal or external to science, is now discussed widely even by those who previously limited themselves to specialized logical and methodological problems of science. In recent years the problems of science, ethics and humanism are being widely discussed by many religious figures and organizations. In particular, the World Council of Churches has produced certain positive findings concerning, among other things, the admissibility or non-admissibility of experimentation on human beings, etc. In general terms, however, it is recognized that scientific discoveries, for instance those in human genetics, and particularly their prospective uses have placed religious ethicists in a difficult position, for in considering many issues they cannot rely on past precedents. Religious ethicists stress the predominance of the moral aspect. Their view is that science and technology are merely means in the hands of men who may use them in pursuing very different objectives; hence the need to subject them and their uses to an ethical control.

In an overall appraisal of the theological approach to the correlation of science and ethics it should be noted that, paradoxical though this may initially seem, its rejection of the fundamental integral unity of science and ethics makes it akin to scientism that juxtaposes cognition and values, science and humanism. In particular, this aspect is quite evident in the activities of Pope John Paul II, who told a conference held to mark the 350th anniversary of Galileo's *Dialogo dei due Massimi Sistemi del Mondo* that while "serious misunderstandings" existed between science and faith, the possibility now exists of overcoming it by recognizing approaches that correspond to different orders of knowledge. To what does this refer? The Pope noted that through experience and subsequent reflection the Church itself has been learning and that today it allegedly understands better than earlier the proper meaning of freedom of research. He expressed the view that there can be no real contradictions between science and faith, since representing "two different orders of knowledge" they contribute to the discovery of "the integral unity of reality" originating in . . . God.

It is characteristic that Pope John Paul II chose the case of Galileo in seeking to "reconcile" science and faith and to stress that today "greater receptivity" of scientists and researchers to spiritual and moral values brings "new approaches" into science and requires a fuller perception of elements of "universal significance". In his opinion, such an approach has also greatly enriched the dialogue between science and the Church. The Pope placed particular emphasis on the point that modern science and technology have become an effective force and an object of socio-economic strategies that are far from neutral in their influence on man's future. In such a context the world of science itself reveals the full scope and significance of its responsibility, and profound moral transformations are called for if we truly wish to place today's scientific and technological resources at the service of man.

The Pope called on scientists, persons he described as enjoying exceptional moral influence, to truly assert what he called science's humanistic and cultural objectives, to protect man and his dignity in the centres where fateful decisions are taken in the sphere of scientific and social planning. He assured them that whenever they strive to achieve man's integral development they would always find the Church to be an ally. The Pope said that for true believers in-depth research could become a highly ethical and spiritual activity and that the world needed minds that could take an integral view of things and that made it possible for knowledge to advance towards an integral humanistic understanding and wisdom. In other words, he noted, scientists' knowledge could become wisdom, a true measure of the entire man. This was Pope John Paul II's final conclusion and appeal to scientists and to their intellectual abilities and ethical responsibility.

As we have seen, this chain of reasoning is by no means trivial, and it makes use of many of the real problems and ethical dilemmas of modern science. Its goal is to consolidate the positions of religion, to "modernize" it and bring it up-to-date with the requirements of modern science which is focusing on man, his moral world and his value orientations.

This is why there is a need for redoubled efforts on the part of scientists, not only in order to positively resolve the problems of the correlation of science and ethics but also so as to oppose religion's claim to a monopoly "right" to discuss and resolve them, or even to participate in these discussions and solutions.

In bringing together the preliminary conclusions of our examination of the various conceptions of the interrelationship between science and ethics I would like to stress that many scientists and thinkers in various countries are seeking their own specific ways to resolve that question. Their specific approaches are determined by the characteristics of these countries' socio-economic structures and cultures. This aspect must be taken into consideration even if we attempt to define these problems in ways that are concrete as well as general.

It has already been noted that in discussing science's ethical problems it is altogether inadequate to view science as merely an aggregate of knowledge that is abstracted from the context of purposeful activities of persons who create this knowledge, communicate it, assimilate it, and use it. It is precisely such a view of science that underlies the proposition concerning its ethical neutrality: if knowledge is viewed as some self-existing reality that is independent of individuals, of their existence and activities, and if, moreover, this particular conception of knowledge is identified with science, then, of course, discussions of ethical aspects of science become pointless.

Conversely, when we view science as a cognitive activity, and as one that is socially organized, it becomes possible to identify an entire spectrum of ethical characteristics of this activity. In that connection one should note that scientific activities continue to be studied primarily as research activities, i.e., as activities aimed at obtaining new knowledge. The latter aspect is indeed a fundamental feature of scientific activities. It is well known, however, that especially today activities of scientists are not limited to research alone and also include teaching, search for information needed to formulate a research project and to discuss appropriate methods, the informing of colleagues as well as disseminating information about scientific achievements. Frequently, scientific workers are also involved in problems of administration, including the administration of research and development. Editing and reviewing others' studies is often still another part of scientists' activities. Finally, their role as consultants or experts in solving various problems is also becoming a common phenomenon. Such a range of responsibilities for scientists is a natural consequence both of the diversity of and interrelations within science's social functions and of the growing complexity of relations within the scientific community that accompany science's development.

In recent years the diversity of forms and types of scientific activity has begun to be studied by such disciplines as the sociology of science and the history of science. That diversity widens the range of problems with which the ethics of science is concerned. Marxists view positively and with respect the position of those progressively thinking scientists who not only defend the interests of scientific research as such but are also aware of science's social responsibilities. Today, when imperialism relies increasingly on science's achievements in serving its self-seeking objectives, that issue acquires a special urgency for Western scientists. Since science now is a complex social activity, comprehensive studies of science must also include ethics and the ethical principles that govern scientists' social activities.

The "ideal" conception of science as something isolated from society, its needs, and its ethical norms, has long vanished into the past. Today the image of the lonely scientist committed solely to the norms of a specialized "craft guild" to which he belongs appears archaic. The illusion that scientific activities are independent of society and its needs, and the corresponding freedom to choose research problems and approaches characteristic of previous stages in science's development, have forever become no more than an attractive memory going back to the early days of science.

The assignment of an absolute value to objective cognition as the sole source of the authentic truth and the scientistic excessive optimism that found its supreme expression in Pierre Laplace's conception of a Universal Mind precluded, as is known, the need for an external—both social and ethical—evaluation of knowledge. Consequently, the existence of any ethics specific to science was out of the question, and ethics was excluded from the field of knowledge as something that did not possess the attribute of objectivity. Ethical evaluation of scientific knowledge was made meaningless and unnecessary, while the problem already posed by Socrates concerning the coincidence of wisdom (knowledge) with virtue (morality) was seen as some kind of paradox for quite a long time. Kant's attempt to curb scientism's ambitions and to show the limits of theoretical reason as well as the need for complementing it by independent moral principles—a "categorical imperative" deriving from usefulness to mankind—appeared to be abstract moralizing.

The collapse of these illusions, which was accelerated by the tragedy of Hiroshima and Nagasaki, made many scientists aware of the new social and moral situation in which they found themselves. In referring to their often bitter and tragic experience it is important to stress that it added substantially different dimensions to discussions of science's socio-ethical problems and posed the question of boundaries between science's "external" and "internal" ethics in a completely new way. It is becoming quite evident that by itself science's "internal" ethics is incapable of offering scientists a reliable compass for avoiding deviations from science's essential strivings and its humanistic mission. In his reminiscences, Max Born noted that while from a personal point of view scientific activities gave him satisfaction and happiness, "in the operation of science and its ethics a change has taken place that makes it impossible to maintain the old ideal of the pursuit of knowledge for its own sake which my generation believed in. We were convinced that this could never lead to any evil since the search for truth was good in itself. That was a beautiful dream from which we were awakened by world events."[1]

Let us recall several episodes from modern history which are sufficiently well-known. The first one is the previously mentioned explosion of American atom bombs over Japanese cities. Unfortunately, in that case the scientists' feeling of social responsibility and their protests failed to prevent the catastrophe, and many of those doing research in nuclear physics at the time felt a deep sense of guilt because of the tragedy. This is how Werner Heisenberg describes, in *Der Teil und das Ganze,*[2] the reaction to this event of Otto Hahn, a German physicist who

---

[1] Max Born, *My Life and My Views,* Charles Scribner's Sons, New York, 1968, pp. 190-191.

[2] Werner Heisenberg, *Der Teil und das Ganze. Gespräche im Umkreis der Atomphysik,* R. Piper & Co. Verlag, Munich, 1971.

had discovered uranium fission together with Fritz Strassmann in 1938. On August 28, 1945 Hahn, Heisenberg and a number of other German scientists were British prisoners. Understandably, they had not been directly involved in the work on the atom bomb that had been taking place in the United States; nor could they somehow influence the decision taken by American authorities. Still, having learned in August 1945 of the bombing of Hiroshima, Heisenberg came to the conclusion that the advances in nuclear physics that had been his life's concern for 25 years had caused the deaths of more than 100,000 persons.

But Otto Hahn, Heisenberg recalls, "was shocked even more than the others. The fission of uranium had been his most important scientific discovery. It had been a decisive and completely unanticipated advance in nuclear technology. And now it was evident that it had brought an entire large city and its population of unarmed civilians most of whom did not consider themselves responsible for the war, to a horrible end. Shocked and shattered, Hahn retired to his room and we were seriously concerned that he might do something to himself."[1]

In current discussions on the ethical problems of science the idea is sometimes put forward that the feeling of guilt on the part of scientists is simply an invention. Hardly so. In many cases feelings of guilt among scientists are a normal human reaction to some of the negative social consequences of scientific and technological progress and a completely natural form through which scientists become aware of their responsibility before mankind. The feelings of guilt that were felt by many Western scientists following the tragedy of Hiroshima and Nagasaki are also a fact of the history of science.

Feelings of guilt, moreover, are not only an objective fact, but may also play a highly constructive role. The example being considered served as a serious lesson that has shown that scientific activities are becoming increasingly closely linked not only with the mankind's wellbeing but also with its very existence. An awareness of that linkage has been a powerful incentive, in that it made many scientists socially and politically active and produced such forms of their social involvement as, for example, the Pugwash movement, initiated by the widely known

[1] *Ibid.*, p. 263.

Russell-Einstein manifesto. As a result, Weisskopf notes, "many basic scientists have made important contributions toward a solution of the arms-control problem. Their activities initiated the discussions that led to the halt of bomb tests."[1] As we know, these matters moved beyond discussions as the issue of the harmful consequences of nuclear tests was subjected to comprehensive scientific study. Significantly, since that time the influence of scientific discoveries on society and the social consequences of scientific and technological development have been an object of continuing close scrutiny on the part of scientists.

The next episode is associated with the ecological movement that emerged in the 1960s, when a concern over a growing pollution of the habitat and a depletion of the planet's natural resources awakened in social consciousness. It is important to note that it is the social responsibility of scientists that provided the initial impetus which led first scientists themselves, and then public opinion generally, to recognize the seriousness of the situation with the environment. Unlike the preceding example, in this instance a responsible attitude of scientists was felt before this situation as a whole became irreversible. And while in the first case only scholars in certain branches of physics were directly involved in the tragic course of events, the ecological movement became essentially universal as regards science, in the sense that it involved representatives of the most varied disciplines in the natural and social sciences and engineering. It is also noteworthy that not only their social concerns but also their professional and specifically scientific interests attract scientists to the movement, as illustrated by the fact that today a substantial volume of scientific research, both applied and basic, is devoted to various aspects of the problem of man and his habitat. We thus see that the social responsibility of scientists is, after all, not an external factor in relation to basic tendencies in the development of science, individual disciplines and lines of research. The 18th session of the UNESCO General Conference adopted on November 20, 1974 a set of special recommendations on the status of scientific workers. Listed among the most important ethical and civic principles that scientific workers in all countries should take

---

[1] Victor F. Weisskopf, *Physics in the Twentieth Century: Selected Essays*, p. 347.

into consideration and be guided with are: the intellectual freedom to seek, express and defend scientific truth as they see it; participation in defining the goals and direction of programmes that they carry out and the methods that must be adopted on the basis of humanitarian, social, and ecological considerations; freedom to express their opinions concerning humanitarian, social or ecological aspects of particular projects and the possibility of withdrawing from such projects if compelled to do so in view of their consequences; a duty to contribute to the development of science, culture and education in their own country, guided not only by the need to solve national problems but also by the United Nations' international ideals.

Today the increased self-awareness of scientists is expressed in various forms. At the 28th regular session of the Pugwash Conference, held in Varna (Bulgaria) on September 1-5, 1978, it was stressed once more that aside from being responsible for their scientific work as individuals, scientists also bear a special responsibility in view of their knowledge, technological potential, and international ties. In their home countries scientists should disseminate accurate information concerning such facts as the consequences of using modern weapons; the consequences of urbanization, of the development of industry, agriculture and social structures; the resources situation as it relates to mankind's future development, and alternatives to the use of such resources, as well as their advantages and shortcomings. The Pugwash Council's declaration states that such activities on the part of scientists, which presuppose the existence of sound relations between them and political leaders and their active involvement in social and political affairs, are needed to overcome misunderstandings, ignorance and hatred and hence to preserve international peace.

Because they reflect a growing sense of social and ethical responsibility on the part of scientists the very fact that all these efforts are taking place is significant in itself. But practice has shown that the diversity of concrete situations in which scientists face such ethical problems is so great that a code of ready-made solutions would hardly be practicable. Incidentally, the same applies to fields other than science as well, for history has not produced universal codes of conduct that would make individuals free of the need to make moral choices in various situations or of responsibility for such choices. That aspect, however, is a specific feature of moral problems, for they have no boundaries and, in a sense, lie outside concrete time. Reflection on these problems has been a spiritual tradition of the human race since ancient times. It is the tradition of its humanistic culture.

Another specific trait of the socio-ethical and moral-humanistic problem area is that it does not produce a unidirectional movement of cognition in which each succeeding solution would complement the preceding one. This is why the moral-humanistic reflections of Socrates or Kant, for example, can often be more relevant to us today than many modern theories.

As in many other phenomena in human culture, the subject's spiritual activity and continuous emotional involvement in situations that often have no similarities or models lends a special meaning to the moral-humanistic problems. In this sphere as well, however, an orientation is possible towards objective value, in the search for which certain laws are revealed.

In that connection one should note above all that from the point of view of concrete research the problems of the ethics of science may appear to be "meta-scientific". Such an approach, however, is unlikely to be productive, for in many cases those problems are interwoven with the very fabric of research proper. To the philosopher, on the other hand, the ethics of science appears as applied ethics which, for the most part, relates to the general foundations of concrete sciences. This deprives it of the status of a philosophical-sociological discipline and reduces its significance, taken in a broader sense, as a necessary and permanent component of all scientific philosophizing.

It would hardly be fruitful to enter more deeply into general definitions and classifications. For in the words of Vladimir Vernadsky "the growth in scientific knowledge in the twentieth century is rapidly eliminating the boundaries between individual sciences. We are increasingly specializing not in particular sciences, but in *problems*."[1] Obviously, this also fully applies to the socio-ethical and humanistic foundations of scientific cognition, i. e., to the problems of the ethics of science.

It was already stressed earlier that the ethics of science can

---

[1] V. I. Vernadsky, *Razmyshleniya naturalista* (Reflections of a Naturalist), Vol. II, Nauka, Moscow, 1977, p. 54.

only develop in the closest contact with the methodology, sociology and history of science and other spheres of comparative studies of science by using the methods and concepts developed in those disciplines and by contributing, in its turn, to the development of more realistic, more comprehensive, and finally more substantial conceptions concerning scientific knowledge and scientific activity. This is why there is no need, in a sense, to develop anew a problem area for the ethics of science, for it has already been developed and is developing further in numerous extremely lively discussions, whose participants inevitably turn to more general problems and to those concerning the interaction of science and society in discussing quite concrete ethical issues that emerge in the development of individual branches of modern science. This is why one of the necessary conditions for a more concrete and thorough study of the ethics of science is the identification and examination of empirical material offered by science's development.

Since the ethics of science cannot pursue the objective of developing prescriptions that apply to all specific cases, it must be a discipline that is, above all, reflective rather than normative. Its basic task is not so much to solve as to identify and clearly formulate the ethical problems and contradictions that emerge in the course of scientific activities, to identify their social roots, and not so much to prescribe to scientists specific standards of behaviour as to provide a critical analysis and substantiation of the ethical norms that scientists actually follow. Ultimately, the above involves a study of the ethical content which is not always explicit but is always present, in one way or another, in scientific activities.

In this respect the ethics of science is similar to the methodology of science which has long abandoned its claim to provide prescriptions for those wishing to make discoveries and now limits itself to analysis and substantiation of the methods and procedures that are applied in science, and also of the premises underlying specific stages in the development of scientific knowledge. The study (but, to repeat, not the invention or prescription) of norms of scientific activity—for instance, that of historically evolving standards of demonstrability and substantiation of knowledge, or of models and paradigms that scientists are guided by—is one of the leading themes in the modern methodology of science.[1] The normative structure of scientific activity, viewed from a specific angle, constitutes the object of study of the ethics of science.

The similarity between science's methodology and its ethics that has been outlined above may be pursued. Like its methodology, the ethics of science is becoming one of the forms of scientists' reflection with regard to the contents, human meaning, and direction of their activities. As in the case of methodology, which has evolved from considering universal ideals of scientific knowledge—established once and for all—and methods for achieving it to analyzing concrete methods of cognition and concrete cognitive situations, the ethics of science can no longer limit itself today to universal propositions relating to all of scientific cognition, such as "science is good", "science is moral", "science is neutral with regard to morality", etc.

Consequently, like questions associated with scientific methodology, those related to the ethics of science are faced by researchers individually and must be resolved by them as questions that concern them personally and without references to "science in general". And just as one should not be misled by the deliberate emphasis of some scientists on their disregard for the philosophical and methodological problems of science, the disregard of *some* scientists for ethical problems does not imply that there does not exist a deep internal linkage between science and ethics.

In that connection one may refer to Heisenberg's view that "for the individual before whom a scientific or technological process has placed an important problem it is not enough to think about that problem alone. He must view its solution as a part of a general course of events that he clearly views as positive inasmuch as he works on such problems at all. If he takes that general interconnection into account he will find it easier to arrive at correct decisions."[2] This, then, is the position of a scientist who seeks to understand the new role of science and of scientific discoveries in human lives and who is not inclined to renounce the responsibilities thus placed upon science.

---

[1] See, for example: *Idealy i normy nauchnogo issledovaniya* (Ideals and Norms of Scientific Research), Byelorussian State Univ. Publishers, Minsk, 1981.

[2] Werner Heisenberg, *op. cit.*, p. 272.

As for the question of freedom of research, note should be made above all of the following. It is widely known that basic research today as a rule calls for joint efforts of large groups of scientists and requires substantial materia resources. This aspect alone imposes unavoidable limitations on the freedom to engage in research. But equally significant is the fact that today science is a fully formed and relatively mature social institution exerting a considerable influence on social life. Accordingly, the idea of unlimited freedom of research, progressive at a time when science struggled against the intellectual predominance of theology, can no longer be accepted without reservations and without considering the social responsibility with which that freedom must be inseparably linked. There is, after all, such a thing as responsible freedom and there is also free irresponsibility.

The juxtaposition of the freedom of research as a demand coming from within scientific activity to social responsibility as something that is imposed from the outside is also based on an excessively narrow view of scientific activities, of their motivations, and of ways in which they are carried out. Science is a search for truth, but the path leading to truth involves scientific and, at the same time, human activities, carried out by man as a whole, rather than by his individual abilities or interests taken in abstract from the individual.

We have considered some of the problems faced by the ethics of science as a field of research in the state of formation. I have sought to show their complexity, intensity, and close linkage both to problems relating to the development of science proper and those relating to its interaction with society. We have seen that particular interpretations of science also have a very substantial influence on the way in which issues pertaining to the ethics of science are formulated and on methods for studying them.

In the methodology of science it is established that particular approaches to science correspond to quite definite schemes for its theoretical-methodological reconstruction, for identifying what is essential and basic and what is secondary and subordinate, its internal and external attributes, etc. As for the ethical problems of science they not only presuppose a certain theoretical reconstruction but also make it necessary to evaluate particular versions of such a reconstruction from a practical point of view. And that is evidence of an unquestionable closeness, perhaps even a certain mutual complementarity, of the methodology and ethics of science which seek to view scientific activity as a multidimensional phenomenon.

At the same time, while the ethical responsibility of scientists is in itself a necessary condition for ruling out possible abuses of the achievements of science it is by no means sufficient. That problem can be solved conclusively only in the course of progressive social development. It would be extremely unwise to exaggerate the potential of scientists in humanizing scientific and technological progress. Yet it would be even more unwise and irresponsible to ignore the current and future impact of an active socio-ethical humanistic position on the part of the scientific community.

There will thus be new problems in and new discussions on the ethics of science, doubtlessly a developing sphere of research. Very probably it will increasingly extend not only to the concrete sciences discussed earlier but also to socio-psychological, social and behavioural sciences, pedagogics, the science of culture, and other disciplines concerned with man, his formation and development. This will not be unexpected, for as a relatively autonomous socio-philosophical discipline the ethics of science can develop only on the basis of studies of concrete situations emerging throughout the entire range of modern sciences and closely associated with their history and the development of human culture as a whole. But that, in turn, means that the range of new problems will increase, as will the difficulties in resolving them scientifically.

This applies especially to the concrete ethical problems in biology, genetics and human psychophysiology, where cognition is governed by certain socio-ethical and humanistic principles. Let us now turn to these problems in order to continue, and perhaps deepen, the philosophical analysis of the subject of "man—science—humanism".

CHAPTER TWO

SOCIO-ETHICAL AND HUMANISTIC PRINCIPLES (REGULATORS) OF THE SCIENTIFIC COGNITION OF MAN

> Man studies life only in order that it may be better. Accordingly, persons who study life are moving the human race forward. But besides these true teachers and benefactors of the human race there have always been and there are today reasoners who abandon the purpose served by reasoning.
>
> *Lev Tolstoy*

That epigraph will offer us a key to establishing the importance of approaches to the cognition of man that are both rigorously scientific and humane as well as the danger in this context of uninformed, even if seemingly scientific, projects and proposals. We shall also elucidate our positive idea concerning the significance in that sphere of strict adherence to the ethics of science and to its humanistic principles.

1. Humanism and Ethics in the Context of Modern Biology; Evolutionary-Genetic Foundations of Ethics? Experimentation on Man: Socio-Ethical Evaluations and Solutions

In continuing our discussion of humanism and the ethics of science we proceed to a number of somewhat different aspects. These are concerned with certain tendencies in modern biology and specifically its "challenge" to humanism and ethics. I have in mind, first, attempts to substantiate ethical-humanistic imperatives and norms on the basis of biological data, and secondly, those achievements and prospects of biological science which entail situations calling for regulation by ethical-humanistic principles.

We have already considered attempts to improve ethics with the help of "rigorous" scientific methods and by using intrascientific ethical norms as a model for reforming ethics at the level of society as a whole. In recent years, if one judges by publications, that problem area has receded into background, for the discussions are now focused on more acute, more concrete, and more urgent questions. Yet the search for a "scientifically substantiated" ethics has somewhat unexpectedly turned in a new direction.

We have seen that the growing interest in the biology of man and the new prospects opening up in that field at the present stage in the revolution in science and technology have intensified biologistic tendencies and sociobiological conceptions. These infer not only man's reason but also his ethical qualities from biological, and in particular evolutionary-genetic foundations, rather than from social conditions. An important role is played in this context by neo-Freudian ideas as well as by biologistic interpretations of ethological conclusions.

Characteristically, in the consideration of the evolutionary-genetic prerequisites of the emergence and expression of certain ethical qualities in man, and in particular of altruism references are made to a number of relevant statements by Charles Darwin—especially those in his *The Descent of Man, and Selection in Relation to Sex.*[1] In a historical context, use is made of the findings and conclusions made by Pyotr Kropotkin on the origins and development of morals and altruism.[2]

Today the biological foundations of ethics, and especially those of its aspects that were formulated under the influence of the ideas of Freud and of ethology (Konrad Lorentz, Robert Ardrey and others) have become the subject of wide discussions in world science, as reflected in a collection entitled *Biology and Ethics* that contains reports and presentations made at a symposium held in London by the Royal Geographical Society.[3] In the introduction, F. J. Ebling, one of the sponsors, defines the most important aspects of the relations between ethical problems

[1] See Chapters 3 and 4 entitled, "Comparison of the Mental Powers of Man and the Lower Animals", and Chapter 5, "On the Development of the Intellectual and Moral Facilities During Primeval and Civilised Times".

[2] P.A. Kropotkin, *Etika* (Ethics), Vol. 1, *Proiskhozhdeniye i razvitiye nravstvennosti* (The Origins and Development of Morality), Petrograd, Moscow, 1922.

[3] *Biology and Ethics. Proceedings of a Symposium Held at the Royal Geographical Society, London, 26 and 27 September 1968.* Ed. by F. J. Ebling, Academic Press, London and New York, 1969.

and modern biology. In summarizing the results of the discussions he notes that Darwin's conception of natural selection has altered irreversibly man's idea of his own origins. That, however, was only the beginning, for both research in human behaviour and other advances in modern science and technology have posed new problems. Today, Ebling continues, it has become clear that in many spheres of the animal world the continuation of a species depends on the survival of social groups rather than of individual specimens. Groups possess two major advantages with regard to selection: they guarantee protection from enemies, and they are more effective in gathering food.[1]

Accordingly, Ebling concludes, the principle of selection begins to operate at the level of groups and within the framework of groups, and joint activities of individual members become factors of immense importance in evolution. This situation results from two phenomena: on the one hand, organisms operate above all jointly rather than as rivals, and secondly, individuals may sacrifice themselves for the sake of the group. It is there that we see the origins of altruism.

Even if we assume that "joint activities", and "altruism" are the terms to be employed to describe behavioural phenomena, are there grounds, nevertheless, for viewing this phenomenon as the emergence of ethical behaviour? In order to answer that question we must first attempt to define ethical behaviour.

In Ebling's view human behaviour is ethical or moral when it is proper and just and thus opposed to improper and unjust behaviour; as for ethical codes, these are systems of everything that is seen as "proper and just" in each individual society. It is the nature of "justice" or "injustice" that represents concrete material for the discipline that we call ethics or moral philosophy. Whatever the philosophical errors and distortions that we commit in seeking to infer or define our value judgements, one important element always remains, namely, that any ethical behaviour must involve a *choice*. There can be no moral sanction or moral approval of actions or acts that are actually inevitable or unavoidable.

In a biological sense the nature of that choice seems clear.

---

[1] William H. Thorpe, *Science, Man and Morals,* Methuen, London, 1965.

That principle emerges from a conflict between mechanisms contributing to the organism's or individual specimen's survival and those that contribute to the survival of the social group as a whole. One could assume that individual stimuli—such as hunger, sexual attraction and fear—possess instinctual components that could be channeled for the benefit of society by means of training and education. Such an approach, however, would not be entirely correct, for all behaviour is conditioned by both genetic and ecological factors, and a moral choice can be formed on the basis of a conflict (or, more precisely, from a conflict) between acquired models of behaviour.

From a philosophical point of view it is interesting that Ebling seeks to compare the view that has just been presented with the corresponding conception of Freud. He writes that the instinct of individual survival and reproduction (the continuation of the genus, of "one's self") may be compared to Freud's "id", and the basic deep-seated attraction to sexual motivation and the striving to satisfy it. Only a part of our ethical behaviour, Ebling adds, passes through our conscious mind.

Turning next to the emergence of ethical behaviour in the course of evolution, Ebling does not lose sight of the fact that examples of altruism can also be observed among birds, that the latter are also able to "learn", etc. Their behaviour, however, is deeply "programmed". To trace the emergence of morality in the course of evolution, he believes, one should consider above all the evolution of mammals, and especially of the primates which are closest to man. He bases that view on the following speculative assumption: the reason why man occupies a unique place and position is because in the course of evolution (or, more precisely, development) his ethical behaviour is conceptualized and codified.

In summarizing Ebling states that the kind of behaviour that is subjected to evolutionary development and is also its outcome is one that, while serving the interests of the group, may also find itself in contradiction with the strivings and desires of the individual. Such behaviour is inherited and, to a certain extent, "built in" the structure of each individual through the "mechanism of consciousness"; it is then reinforced and maintained in society through an entire system of devices that range from neurohumoral sanctions to threats in the name of law. Ethical conceptions, according to Ebling, are not simply prescriptions originating in society that are "addressed" to individuals, since individuals themselves also give these prescriptions to society.

Thus, in seeking to establish a relation between man's biological nature, seen as an element of the animal world, and ethical norms, Ebling does not arrive at any consistent conclusion. He merely observes that ethics possesses biological, evolutionary foundations.

An attempt to make the latter proposition more concrete is made by M. R. A. Chance in an article entitled "Towards the Biological Definition of Ethics". Chance considers that behaviour is a link between the organism and its environment, and that its study will give us a greater understanding of the process noted by Darwin. But no one succeeded in doing this until the appearance of ethology as the most recent product of behavioural science. Next the author provides a logical justification for considering ethological aspects in studying organisms and their behaviour, and, in his opinion, that is where the problem of ethics arises. Ethics, he believes, is a way that is recorded in language for achieving desirable changes in behaviour. The author recommends that principles of ethology be used in selecting leaders both in the political sphere and the economy.

In his article "The Development of Moral Attitudes and Behaviour", B. M. Foss observes that in the case of animals that live in groups, as is the case with man, the evolutionary value of behaviour derives from what is important for the survival of individuals and moves towards what is important for the survival of groups. Naturally one expects the conclusion that on a broader scale such a survival has to depend on the evolution of the structure of the groups themselves, although it is also possible that selection will have produced new individual characteristics. For example, it is entirely conceivable that changes resulting in increased submissiveness (such as an individual's wish to be led or guided by someone) may enhance the viability of the group, even though submissiveness as such may be to a certain "nonrigourous" extent harmful to the individual himself. Yet while in the animal world altruistic acts are rooted in instincts and their mechanisms, in the case of man they are the outcome of training.

I have considered the ideas expressed in *Biology and Ethics* at such length because they subsequently served as a point of departure for many similar views. In particular, this is true of the ideas that are defended by Edward Wilson in his *Sociobiology* and a number of other studies. Altruism, in his opinion, should be viewed from the point of view of the species. Wilson does not agree with the idea that the existence of altruism in the animal world contradicts the conception that evolution is based on natural selection. To the question of how altruism, which by its very meaning contradicts individual adaptability, could emerge and develop in the process of natural selection his answer is that this is explained by kinship. If genes producing altruism are found in two organisms as a result of their common origins, and if acts of altruism increase and multiply the joint contribution of these genes to the next generation, then a propensity for altruism will extend to the entire gene pool. Next Wilson seeks to substantiate the proposition that in the complex process of evolution in the case of highly developed creatures the hypothalamic-limbic system "places into circulation", as it were, an effective mix of principles encouraging individual survival, reproduction, and altruism. Complex relations develop between the individual and the species that result in the preservation of the species and in enhancing its survivability.

The basis for such an approach to the origins of altruism was formulated in 1964 by William Hamilton, a British biologist who viewed altruism as a quality developing in the process of biological evolution in order to help individuals to disseminate their genes. Altruism is thus defined essentially as genetic egoism. In particular, on this basis Hamilton seeks to explain the social life of insects. In all the genera of ants, bees, and wasps, daughters of a single queen have, on the average, three-fourths of their genes in common. Hamilton assumes that since these daughters are linked to one another by closer kinship ties than their own progeny their genetic interests require not reproducing but helping the queen which produces additional daughters. Thus females appear that function within their community as infertile workers for reasons of genetic egoism.

Robert Trivers, who is a biologist at Harvard University, developed these ideas further by proposing a conception of reciprocal altruism that would explain, in particular, the fact that certain birds produce danger warning signals for the entire flock even when the flock does not contain their own fledglings or close relatives. This kind of behaviour, he asserts, is characteristic of all organisms. Therefore, the statement that humans are the only species in which altruism is not of genetic origin is incorrect.

That conception has become the ultimate foundation of sociobiology, and Wilson extends it not only to altruism but to other ethical phenomena as well. In his opinion, both scientists and humanists should consider together the possibility that "the time has come for ethics to be removed temporarily from the hands of the philosophers and biologicized."[1] Contemporary ethical philosophy, he states, consists of certain fragmentary attempts to conceptualize that science. The first one of these he describes as "ethical intuitionism", which consists, in his opinion, in a conviction that human consciousness, reason, or mind possesses a capacity for directly distinguishing between what is good and what is bad, good and evil, as well as an ability to formalize the knowledge thus acquired in a logical system and then transform that system into rules of social actions and of social activities of both individuals and society as a whole. He believes that a typical example of such an approach is the theory of social contract. Among modern philosophers, sociologists, and anthropologists, Wilson mentions John Rawls who has transformed that idea, in his opinion, into a rigorous philosophical system.

That conception's Achilles' heel is, in Wilson's view, the fact that it rests on the brain's emotive judgements (referring essentially to the brain as the organ of consciousness), as if the brain itself were some "black box". In fact, Wilson states, the brain has long ceased to be a mystery and is currently being widely studied from a variety of points of view. The human genotype and ecosystem cannot be excluded from ethics. In any event, he concludes, exhaustive studies of the neuromechanism of ethical judgements are highly desirable and are already taking place.

Wilson considers what is known as ethical behaviourism as another attempt to conceptualize ethics. Its basic proposition is, in effect, that moral behaviour originates entirely in learning, and that, aside from the process of learning, there also exists a certain dominant mechanism, some essential conditioning element the role of which in all such conceptions is played by society. In

---

[1] Edward O. Wilson, *Sociobiology,* p. 562.

short, there is a determining principle which enables children to fully assimilate or internalize a particular society's behavioural norms.

Opposing these views, according to Wilson, is the evolutionary-genetic conception. "Ethical philosophers intuit the deontological canons of morality by consulting the emotive centers of their own hypothalamic-limbic system... Only by interpreting the activity of the emotive centers as a biological adaptation can the meaning of the canons be deciphered."[1] Wilson then elaborates on this assertion in the language of biological research. His point is that as the philosophical discipline closest to biology ethics cannot be constructed through logical means alone, for it is associated with interpreting human behaviour, a phenomenon rooted in the biological evolution of man and primates. In the case of such phenomena as altruism, moreover, their origins can even be traced to the history of invertebrates.

This is why, Wilson states, ethics as a science must be removed from the hands of the philosophers and biologized. It remains unclear, however, why it is not possible to study and formulate many elements of ethics already at this stage, when it has been objectively developed by society (as the sum total of norms of behaviour) and by scholars (as philosophical and world-view systems), and why biological-genetic studies cannot simply serve to complement philosophical studies or even as an autonomous approach to problems of ethics, especially its genesis. It is obvious that one-sided biologization of social phenomena cannot shed light on the evolutionary-genetic foundations of egoism and altruism, which have been studied quite successfully in recent years; more generally, it does not even make understandable the role and significance of the biological component in the formation of human behaviour which we evaluate in terms of ethical categories. Such categorical conclusions as those made by Michael Ruse, for example, to the effect that it is more than probable that we inherit ethics and moral feelings in accordance with biological canons are hardly convincing. Nor is it surprising that the strongest objections in *Sociobiology: Beyond Nature/Nurture?* are addressed explicitly to the sociobiological interpretation of ethics, of values more generally, of politics, and so on.

---

[1] *Ibid.*, p. 563.

If the method of reduction of ethics to its biological component is applied, many negative traits of human behaviour and such social phenomena as wars are perceived as being "inborn" (and are thus partly justified as "natural" and "eternal"). Still, at least in terms of subjective intentions, a number of scientists are seeking to place that method at the service of humanism and of human progress. This tendency is seen most clearly in the works of Julian Huxley and his conception of "evolutionary humanism".[1] Essentially it is based on an assertion that man's striving towards education, culture, and science is conditioned by evolutionary-genetic factors. It represents continuing development and perfecting of the "human nature" which, unlike Lorenz and Ardrey, Huxley interprets in an altruistic spirit. This is why Huxley views the future optimistically, even though he does this from abstract-utopian positions.

I believe, however, that for a number of reasons Huxley's views merit closer examination, for he has anticipated a number of modern conceptions of the future in the context of global problems and of a "new humanism" concerning which much has been written, for example, by the participants in the Club of Rome, unfortunately without reference to Huxley.

As the question has been dealt with previously, let us only note the main point, namely, the general methodological limitations characteristic of the "evolutionary-genetic" ethics and its extreme conclusions, including those of Lorenz, Ardrey, Huxley, and also those who are now joining forces under the banner of sociobiology.

Similar ideas concerning the genetic origins of altruism are expressed in *The Selfish Gene* by Richard Dawkins.[2] He writes that like other animals human beings are machines created by their genes. In a world of competition these genes could survive only through their ruthless egoism. Since genes are egoistic one should expect that our individual behaviour, too, will be equally egoistic. Accordingly, types of behaviour that initially seem to be altruistic emerge only because in certain specific situations

---

[1] Julian Huxley, *The Human Crisis,* University of Washington Press, Seattle, 1963; *idem., Essays of a Humanist,* Penguin Books, London, 1966.

[2] Richard Dawkins, *The Selfish Gene,* Oxford University Press, Oxford, 1977.

altruism serves a gene's egoistic objectives. Genes are not only egoistic but also a "herd-oriented", they are merciless in rivalry, and can be "happy", "loyal", "ruthless", etc. In short, as Stent rightly notes, Dawkins writes about genes as if they formulated conscious objectives.

In the last chapter, entitled "Memes: the new replicators", Dawkins begins to deny his own earlier assertions. He states that man is not a simple gene machine but a "unique" species, and that the ideas of evolution are not directly applicable to his social behaviour. The point is that man is a product not only of organic but also of a cultural evolution, in the course of which the basic unit that transmits the ability to develop and assimilate ideas ,the "meme", plays a rôle that is similar to that of genes in organic evolution. The "memes' " success in cultural evolution, however, is also attributable to their "egoism". In this connection, the geneticist Stent, whose position with regard to that issue seems to us to be correct, made a point that is well taken. In a review of Dawkins' study he observed that usually the concepts of "egoism" and "altruism" are associated with the concept of moral value, i.e., altruism refers to a person's respect for the interests of others, while egoism refers to their disregard. In Dawkins' case, however, both concepts lie outside the conceptions of good and evil, since they refer entirely to conditions of survival rather than to respect for others' interests. To Dawkins, the fact that some animals kill and eat others would illustrate "egoism", while a hen's protection of her chicks would illustrate "altruism". Both lie outside the ethical sphere. It would seem that he needed the concept of "egoism" to replace what Darwin called "adaptability".

Dawkins offers his own definition of a gene as any part of chromosome material that is potentially preserved over a number of generations sufficient for it to serve as a unit of natural selection. But he admits that no such unit exists in the sense that, depending on particular biological situations and the harshness of conditions within which selection takes place, any part —large or small—of the genomic material may be selected. Accordingly, in Stent's opinion, that distorted definition replaces the established concept of genetics by one that is diffuse and heuristically useless. Dawkins himself explains the use of the word "gene" by his wish to find a catchy title for his book.

Such conceptions, of course, call for criticism since their analytical *methodology* itself is faulty. What makes it fallacious, contrary to what is sometimes asserted, is not that an attempt is made to study the role of natural-biological factors in man's social behaviour, including his ethical qualities. Engels and Lenin spoke of the importance of such studies concerning man's thinking and consciousness. From such a point of view evolutionary-genetic and ethological data are extremely important. Engels, in particular, wrote: "It is, however, inherent in the descent of man from the animal world that he can never entirely rid himself of the beast, so that it can always be only a question of more or less, of a difference in the degree of bestiality or of humanity."[1] But in the works of many ethologists, neo-Freudians, sociobiologists, and those who rely on their conclusions concerning the origins of ethical values, the significance of biological factors is exaggerated and sometimes made absolute, while the importance of social factors is played down. And yet, according to Lenin, "nothing is easier than to tack an 'energeticist' or 'biologico-sociological' label on to such phenomena as crises, revolutions, the class struggle and so forth; but neither is there anything more sterile, more scholastic and lifeless than such an occupation."[2]

The position of "extreme genetic determinism", that is of social-biologism, in considering man's ethical qualities is found in its extreme form in Wilson and Lumsden's *Genes, Mind and Culture: the Coevolutionary Process,* which I have analyzed elsewhere. There it is not only ethics but human culture generally that is determined by some "culture genes" that are said to exist in the human genotype. Naturally, that position has been criticized not only by Marxists but also by many Western researchers who do not adhere to Marxist positions, for instance Lawrence M. Hinman, whose article is entitled "The Ambiguity and Limits of a Sociobiological Ethic".[3] The point, how-

---

[1] Frederick Engels, *Anti-Dühring,* Progress Publishers, Moscow, 1975, p. 118.

[2] V. I. Lenin, "Materialism and Empirio-Criticism", *op. cit.*, p. 328.

[3] Lawrence M. Hinman, "The Ambiquity and Limits of a Sociobiological Ethic", in: *International Philosophical Quarterly,* Vol. XXIII, No. 1, Issue No. 89, March 1983, pp. 77-89.

ever, concerns not merely the ambiguity and limitations of a sociobiological interpretation of ethics. It is generally fruitless to consider any sphere of human activity and its corresponding values outside their general social context. Otherwise, values are viewed as extensions, as it were, of purely biological evolutionary-genetic qualities since they possess a common "naturalistic" basis.

In Marxist literature the methodological positions of naturalism and of social-biologism, particularly those concerning the evolutionary-genetic basis of man's ethical values and especially altruism, have met with well-founded criticism.[1] At the same time, attempts are being made to overcome the extreme positions of naturalism and social-biologism by emphasizing the significance of evolutionary-genetic prerequisites of certain ethical qualities in man and in particular of altruism. It should be noted, however, that a number of inconsistencies and contradictions have emerged in many approaches to that problem. For example, in a discussion of the socio-ethical problems of human genetics, Vladimir Efroimson, a Soviet scientist, expressed the view that if one examines closely the specific characteristics of the natural selection to which the human race was subjected during its formation, it will be found that it is precisely natural selection, the harshest of nature's laws, that produced altruistic emotions of exceptional intensity and resilience in man.[2] Unlike sociobiologists, however, he by no means believes that it is up to geneticists to have the last say on the origins of human morals and humanism. What is important is that philosophers do not miss the point that ethics emerged under the influence of not only social but also biological factors.

Efroimson subsequently elaborated his ideas in an article entitled "Altruism's Family Tree", the basic propositions of which were endorsed in general terms by Boris Astaurov, a prominent

---

[1] See especially A. F. Shishkin's article *"Etologiya i etika"* (Ethology and Ethics), in: *Voprosy filosofii,* No. 9, 1974. These ideas, generalized and developed further by the author, were published posthumously in *Chelovecheskaya priroda i nravstvennost. Istoriko-kriticheski ocherk* (Human Nature and Morality. A Historical Critique), Mysl, Moscow, 1979.

[2] See: *Voprosy filosofii,* Nos. 7, 8, 1970.

Soviet geneticist.[1] The principal point is that with his potential reason and moral qualities man is not born as some kind of a "blank sheet of paper" on which his environment and training can record any text. According to Efroimson, there are grounds for believing that man's eternal striving towards justice, heroic deeds and self-denial derived from his hereditary nature. He stresses, moreover, that "the proposed idea in no way rejects the role of the social environment and education in the formation of personal ethical principles. There can be no doubt that any organism, and especially man, with all the features of his psychology, behaviour, and ethics, is a product of the environment."[2] The author specifies that his attention is focused on the role of the hereditary mechanism in the formation of ethical principles, leaving aside for the time being the role of social conditions and the influence of labour on the formation and development of these principles. Still, the problem remains unresolved, for I believe that it is impossible to study the formation and development of man's ethical qualities outside the social context. Without taking into consideration the dialectics of the social and biological components in man, or the mediation and transformation of the biological component through mankind's socio-historical experience, it is not possible to properly demonstrate the actual role of the evolutionary-genetic prerequisites of man's ethical qualities, and in particular of altruism.

Unfortunately, that prevalence of the social component over the biological one, the latter's subordination and its qualitative transformation were not fully present either in Astaurov's article " 'Homo Sapiens et Humanus': Man with a Capital Letter and the Evolution of the Genetics of Altruism", which contains a wealth of factual material drawn from biological history. Astaurov did consider the overcoming of the dichotomy between "education" (the environment) and "nature" (heredity), the leading role of the social environment, the determining influence of social factors and principles in the evolution of the human race and its ethics, as well as the point that "social systems and

---

[1] See. V. Efroimson, "Rodoslovnaya altruizma: etika s pozitsii evolyutsionnoi genetiki cheloveka" (Altruism's Family Tree: Ethics from the Point of View of Man's Evolutionary Genetics), *Novy mir,* No. 10, 1971.

[2] *Ibid.*, p. 194.

processes bearing on the retransmission of the vastly increasing and rapidly changing capital of human knowledge, customs, and conceptions have come to play a role, in the social phase of the human race's evolution, that overtakes by far that of relatively slow transformations in the inherited gene pool, but without dispensing with the fundamental importance of biological laws."[1]

All this produces new discussions. But the major factor is a clear awareness of the need for further positive studies of the evolutionary-genetic prerequisites of man's ethical qualities on the basis of a Marxist understanding of man's social essence and with due consideration of the dialectics of social and biological factors. Such an approach will, I believe, produce new discoveries and generalizations that will reveal more about humanism and ethics in the context of modern biology.

The above, however, is only one side of the issue. The other is an active influence of socio-ethical and humanistic principles themselves on the course of the biological cognition of man. It is there that the regulative role of these principles is revealed in its sharpest form. Bentley Glass, a well-known American biologist, notes that "the gravest ethical problems of our present day arise from the advances which are being made in science at the present time, not in the nuclear science only, but especially in the biological sciences... I insist that philosophers of science would do well to think a great deal more in the future about ethical problems than about logical problems."[2] This refers to the total danger for man and the human race of military applications of biological research, the potential consequences of the use of psycho-pharmacological substances, organ transplants, etc.

This is why today there is a growing insistence on the need for more rigorous socio-ethical principles limiting experimentation

[1] B. L. Astaurov, " 'Homo sapiens et humanus': chelovek s bolshoi bukvy i evolyutsionnaya genetika altruizma. Po povodu stati V. Efroimsona ob evolyutsionno-geneticheskikh osnovakh etiki" ('Homo Sapiens et Humanus': Man with a Capital Letter and the Evolution of the Genetics of Altruism. Concerning V. P. Efroimson's article on the evolutionary-genetic foundations of ethics), *Novy mir,* No. 10, 1971, p. 223.

[2] Bentley Glass, "A Practitioner's View of the Biological Sciences", in: *Mind, Science, and History.* Ed. by H. E. Kiefer and M. K. Munitz, State University of New York, New York, 1970, p. 208.

on man, and, even more importantly, for stricter social and ethical controls over adherence to these principles. These require an unconditional recognition of the uniqueness and freedom of individuals which would be guaranteed by moral and legislative norms as well as actions on the part of society as a whole. But there is also a need for free development of cognition and experimentation on the basis of these norms which are equally important for everyone and must, accordingly, be comprehensively encouraged by society—both morally and through legislation. Of course, any experimentation on human beings is a partial infringement of man's inalienable rights and freedoms; it may, however, be limited to such an extent as to correspond to society's system of moral and other values and be an outcome of freely taken decisions.

To a substantial degree these necessary limitations are determined by the type of society that adopts them and carries them out, and by the way in which the principles of humanism and man himself are perceived. Today, however, it is the view of many scientists that independently of existing differences, a mutually acceptable system of international agreement is needed for regulating man's biological (e.g., genetic medical, etc.) knowledge. There is a need for a strict ethical evaluation of both the theory and practice of cognition and for forms of social control based on national and international legislation. Most scientists agree that humanistic principles must be followed ırom the very beginning of preparations for an experiment, from the outset, and that they then cannot be ignored for the sake of possibly significant discoveries that would be made *post-hoc*. In experiments with man, as in all cases, the end never justifies the means.

These problems are considered at length by experts in *Experiments with Man.*[1] First the grounds are established for experiments of various types. Both in medical practice and in biomedical research there are different types of experiments which often alternate, each of them producing moral problems specific to it only. Participants in the discussion have identified four basic types (other definitions are also possible) in order to make precise ethical distinctions among thme, without either approving or condemning any one of them at that particular stage.

[1] *Experiments with Man. Report of an Ecumenical Consultation,* Ed. by Hans-Ruedi Weber, Friendship Press, New York, 1969.

Type 1. Both physiologically and psychologically each human being is unique. This is why even a conservative treatment of a patient will to some extent be an experiment that is associated with risk. The number of variables is always so large that a measure of doubt must always be attached to its outcome. In that sense medicine has always been an experimental science. In effect, it has proceeded from one working hypothesis to another and from one deductive conclusion to another; these were arrived at following close observation of largely uncontrollable and random clinical events, their close study, and verification of consequences.

Type 2. Following the application in medicine of methods of experimental science there arises a need for a planned collection of data lending themselves to verification and to expression in quantitative terms. Doctors also become research scientists while patients become objects of experiments. While restoring a patient's health continues to be the basic objective, data are also collected in the course of treatment that help in testing and improving hypotheses, therapeutic methods and medicines, and that provide a basis for new ideas.

Type 3. A different procedure for performing experiments is developed for basic clinical research. Its basic objective is to increase existing knowledge of physiological, biological and psychological functions generally, independently of whether this will bring direct and immediate therapeutic results.

Type 4. New advances in human biology and their application to social problems are increasing the complexity of ethical problems. A new line in biomedical research is the manipulation of genetic material on the basis of knowledge concerning the chemistry of genes and other discoveries in molecular biology. In the future such knowledge could be used for overcoming genetic defects or organizing genetic changes corresponding to predesigned models in serving the interests of society. Already today feasible procedures can be imagined for changing human personalities with the help of neurosurgery. In the future the same will apply to neuropharmacology. Ethical principles associated with such research will greatly differ from the ethical guideliness of the first three types. In such cases society and the researcher must act as guarantors of the preservation of future generations. This requires that admissible limits be established to the manipulation of the basic genetic material of man as a biological species.[1]

In such a context the authors of *Experiments with Man* arrive at a very interesting, yet philosophically debatable, critical analysis of basic assumptions underlying biomedical research. In their opinion, since medical practice is partly based on the natural sciences, it must share with applications of science to technology a common set of methods and basic postulates, namely, that matter is cognizable through observation and experiments; that it is quantifiable in terms of logic and mathematics; that these objectives are attainable through experiments; and that it is possible to apply scientific knowledge to technology in order to improve the situation of man.

The authors consider that in the case of medical practice and especially medical ethics the following fundamental definitions are essential:

—man is a unique phenomenon within the Universe;

—man is able to reflect and enter into social relations; he is able to give expression to his freedom by agreeing or refusing to serve as an object of experiments; he is able to arrive at moral judgements in relation to himself and assume responsibility for his actions;

—being what he is, in becoming an object of scientific research and experiments, man requires a special attitude to himself and defines limitations for the researcher or physician that by their nature fall outside the framework of science.

The formulation of these assumptions does not imply their approval in their present form. Their critical consideration in the light of scientific experience points to the need for changes. In particular one cannot deny that experiments with man can produce genuine scientific knowledge, mainly descriptive and non-quantifiable.

Beyond this, confidence in techniques, taken in its oversimplified form, is subject to reservations in the light of the following:

---

[1] In that connection it should be recalled that, in the opinion of a number of scientists, even if one per cent of undesirable pathological genes will be replaced in all persons by normal ones, combinations of the remaining 99 per cent (50,000) can fully assure the kind of practically infinite diversity of individual heredity programmes that exists today.

—the techniques depend on economic, social, and, of course, moral inputs that are not always present;

—each major technical decision, even when made with good intentions, can produce new problems that are even more complex;

—such decisions can be taken in serving both good and evil interests.

Because of the ambiguity of the situation the authors believe that it calls for the establishment of ethical norms for biomedical research and practical activities. In that connection, in their opinion, the following conflicts and dilemmas must be taken into consideration:

(a) there exists a gap between healthy scientific progress and public expectations. The rate of scientific progress is relatively modest; for prolonged periods of time, it is limited to the sphere of "pure research" that has no immediate practical applications. Moreover, even when practical results begin to emerge true science seeks to thoroughly verify relevant hypotheses with the help of experiments that are thorougly prepared, controlled and well-considered. The public, on the other hand, (regardless of whether it is the population as a whole, governments, journalists, or organizations financing research) thirsts for major scientific discoveries and technological advances that would make it possible to solve existing problems and cure illnesses quickly, conclusively, and effectively. The authors sharply pose the following issue: how can one avoid the consequences of such pressures from an impatient public that could force researchers to either withdraw to an "ivory tower" or else consent to a "bad science" that jumps from unproven or semi-proven hypotheses to stating them as facts?

That dilemma is made both sharper and highly personal in relations between medical researchers and patients. When should one apply or cease to apply a medicines or a therapeutic method if they were never subjected to thorough testing? When is it permissible to employ new medicines and methods of treatment that have not been fully tested if one wishes to avoid an excessive gap between the current level of biomedical research and traditional medical practice? To what extent should patients be informed of the ambiguous nature of medicines or methods that are employed and of the corresponding uncertainty?

New discoveries produce new problems. There are widely known examples associated with biomedical research:

—to a significant extent, rapid population growth and over-population as one of its negative consequences have resulted from medical advances in struggling against child mortality;

—the death of old people who are no longer able to lead a meaningful life, is postponed by sustaining their vital activities with the help of strong medicines. This raises not so much a problem of "euthanasia" (which is currently generally taken to mean killing by administering a lethal dose of some medicine) as of "orthonasia" (the use of pain-relieving medicines that allow the patient to die peacefully, without resorting to medicines used to artificially sustain vital functions);

—the unanticipated discovery that the wide use of certain effective insecticides entails harmful long-term consequences;

—interference in the "survival of the fittest" (as a characteristic trait of natural selection). In the authors' view, while such an intervention is desirable as an expression of love for the weak, it may also produce a deterioration of man's genetic material and thus pose a threat to the future of humankind. (Today, a number of prominent scientists are disproving the widely held view that human genetic material has indeed been deteriorating over a prolonged period of time).

These and other uncertainties associated with biomedical research raise the issue of what should be done with already existing dangerous knowledge and with similar knowledge about to emerge. How should we respond to the dilemma between assured or expected short-term results that are positive for individual patients—and the route that is still unexplored and may lead to harmful consequences in the long term? While biomedical research seeks to benefit man, it often produces harm. Should one consider that this is one of its inherent attributes? Should biomedical research proceed, making new discoveries, or are there limits to such research and its autonomy?

The very situation of scientists engaged in biomedical research is already ambiguous and contradictory. Their motivations may be curiosity and a striving for truth, a wish to improve the human race's situation, a desire for recognition or possibly some less lofty motives. One must also recognize that researchers work in a state of extreme tension because of the expectations that public opinion and financing institutions attach to their work and because of practices guiding appointments to particular posts in research institutions and institutes of higher learning. An important role is also played by the researcher's awareness that no matter how good his own intentions may be his findings will not remain under his control. They may be used for purposes of bacteriological warfare or for irresponsible genetic manipulations. Researchers who find themselves in such contradictory and ambiguous situations generally have few opportunities, if any, for discussing with faculty or laboratory colleagues their motivations, the problems of choice that they confront, and possible consequences of their scientific work. A discussion of such problems cannot be delegated to scientific workers concerned with problems of ethics, who would discuss them primarily privately over a cup of tea outside working hours. How can such discussions go encouraged already during training in universities and how can this be done in such a way that these problems then become an integral element of biomedical research?

As we see, this poses many urgent and important issues that appear to require clear answers from scientists throughout the world, independently of their world outlook and philosophical or social position. But the sensationalism with which many biological and medical discoveries have been announced in the West in recent years does not contribute to this. Suffice it to recall, for example, the many speculations that followed experiments in transplanting artificially fertilized human egg cells and the successful birth of the first "test-tube" baby. And yet, of course, one cannot ignore the real and occasionally very complex problems that arise in that connection and that require, in the opinion of many scientists, a significant "shift in thinking" as well as new ethical and legislative guidelines that would clearly define the boundaries between permissible and impermissible interference into the sphere of family and intimate relations.

In this context, in addition to ethical regulators an important role is played by legal and legislative decisions. The laws of many countries, including the USSR, permit clinical experiments on man. It must, however, be clearly established when, how often, to what limits, and with what degree of risk such experiments are permissible. Of course, any scientistic manipulation that is incompatible with the ethical principles of humanism and the sovereignty of individuals should be made totally inadmissible. How this can be determined in specific cases is apparently a matter that cannot be decided by the scientific community alone. It is with good reason that discussions develop more and more often and with growing insistence concerning the elaboration of global criteria with regard to experiments with man that could be embodied in universal codes.

Examples are provided by the Nuremberg Code (1947), the International Code of Medical Ethics (1949), the Geneva Declaration (1948), and the Helsinki Declaration (1964) which contain recommendations to medical doctors engaging in clinical research and ethical principles for experimenting on human beings as well as objective and subjective criteria of ethical permissibility or impermissibility of performing particular experiments with man. Of particular relevance to the situation today are the addenda to the Helsinki Declaration that were adopted by a colloquim held in Switzerland on the subject of Experiments with Man, as well as Articles 6 and 7 of the International Covenant on Civil and Political Rights, adopted by the United Nations in 1966, and other United Nations documents.

An important role in the particular sphere of science is played by the World Health Organization (WHO), the World Medical Association (WMA) and the Council for International Organizations of Medical Sciences (CIOMS), which have produced a number of documents on medical ethics. A major positive contribution was made by a conference on the subject of Protection of Human Rights in the Light of Scientific and Technological Progress in Biology and Medicine, jointly organized by CIOMS and WHO in Geneva in the fall of 1973. Its participants stressed the need for a number of specific limitations and restrictions, in particular in the field of genetics and psychosurgery, in order to protect the rights, dignity and freedom of man. A document was adopted in which the limits and forms of experimenting with man are defined.[1]

In compliance with a resolution of the United Nations Gen-

---

[1] *Protection of Human Rights. In the Light of Scientific and Technological Progress in Biology and Medicine.* Ed. by Simon Btesh, WHO, Geneva, 1974.

eral Assembly on protecting the human person and physical and intellectual inviolability in the light of progress in biology, medicine and biochemistry, the World Health Organization prepared a special document on human rights in connection with progress in biology and medicine, which addresses, in particular, issues relating to experiments on man and the organ and tissue transplants. The document analyzes various approaches and conditions under which such operations may be scientifically and ethically justified. Other problems that are discussed include the beginnings of life, artificial interruption of pregnancies, the use of human embryos in research, establishing the moment of death, psychosurgery, etc.

At the present time there exist international guidelines for biomedical research involving human subjects that were first considered by the Council for International Organizations of Medical Sciences at the XVth Round Table Conference (Manila, Philippines, 1981) and were approved by the WHO Global Advisory Committee on Medical Research at a session held in Geneva (October, 1981) for subsequent transmission to health ministries, medical departments and institutes, and other interested institutions. In an article entitled "Human Experimentation and Medical Ethics", Frank Gutteridge explains the reasons that led to the preparation of the recommendations.[1] He notes that while the basic ethical principles that are recommended as guidelines in carrying out biomedical research involving human subjects are contained in the World Medical Association's Helsinki Declaration, as revised in 1975 by the 29th World Medical Assembly in Tokyo, it is not a simple task to introduce these principles in practice. The growth of medical research especially in the developing countries, has created a need to study the possibility of using the Helsinki principles with due consideration of existing legal provisions and administrative regulations. This is needed to ensure adequate protection of human rights and of man's social wellbeing when participating in medical research.

The basic elements of ethical regulation of biomedical research

---

[1] Frank Gutteridge, "Human Experimentation and Medical Ethics. International guidelines for biomedical research involving human subjects", in: *WHO Chronicle,* Vol. 35, No. 6, 1981, pp. 212-215.

involving human beings include: first, a conscious agreement on the part of the subject, and secondly, an analysis of the proposed research procedure (including the procedure for obtaining the consent of the subject) by a body that is sufficiently qualified and independent of the researchers.

Four basic questions emerged to which complete answers have not yet been provided:

—the concept of conscious consent and its implementation;

—the ethical aspects of biomedical research;

—procedures for analyzing the ethical aspects of such research involving human beings;

—the protection of the interests of persons participating in experiments.

The international recommendations that were produced in the form of general comment and supplemented by specific directives are an attempt to provide answers to these and other questions relating to research involving human subjects, including compensation for damage incurred as a result of research. For example, with regard to conscious consent required clarifications are made for cases of legal incompetence (e. g., with regard to physically or mentally defective persons or persons below legal age) of the subject. In such cases it is required that the question of their participation be examined by their legal guardians or by a body that is independent of researchers. The procedure for analyzing ethical aspects requires that special committees on ethics be established, consisting of independent colleagues of proposed researchers and of persons from other occupational groups. As for the protection of persons participating in research activities, the criteria in considering requests should be especially strict in the case of children, pregnant women and nursing mothers, persons who are mentally ill or retarded, representatives of developing groups of the population who are unfamiliar with modern clinical conceptions, and in cases of any non-therapeutic interventions.

With regard to childen, the recommendations note that it is impermissible to use children for research activities that can be performed equally well on adults. Research of this type is needed, however, in studying child diseases. Similarly, research should not be performed on persons who are mentally ill or retarded whenever it is possible to perform it with equal chances of success on mentally normal adults—except for research on the causes and treatment methods of mental illnesses.

Pregnant women and nursing mothers should never be subjected to non-therapeutic research other than in studies of pregnancy or breast feeding. Therapeutic research is permissible only for the purposes of improving the health of mothers without detriment to that of the fetus or infant, or for improving children's vital processes or contributing to their healthy development.

Accordingly, modern science's socio-ethical and humanistic problems have gone beyond the sphere of abstract moralizing and are becoming a practical issue of vital urgency for all mankind. Modern science also faces urgent problems in medical ethics and deontology. The latter, however, call for separate consideration. I would only note that in this country deontology is now receiving serious and continuing attention, for instance, in the USSR Academy of Medical Sciences.[1] We know that the ethical problems of biomedical research are widely discussed in the West. While we may give different interpretations to certain facts and condemn the sensationalism with which these discussions are sometimes associated, it should be taken into account that in the United States, for example, there is not only an intensive development of biomedical research as such but there is also serious work on a range of problems associated with the controversial notion of "bioethics". That concept gained currency following the publication in 1969 of Van R. Potter's *Bioethics: Bridge to the Future*.[2] Many experts throughout the world participate in the activities of numerous international organizations addressing the ethical problems of biomedical research, including those of experiments with man.

One should also note that studies in the philosophy of science, too, are now placing a greater emphasis on ethical problems. At the 7th International Congress of Logic, Methodology and Philosophy of Science (held in Salzburg, Austria, in 1983) a special section was organized to consider not only the basic principles of the ethics of science, but also problems specific to

[1] See, for example: *Aktualnye etiko-deontologicheskiye voprosy sovremennoi meditsiny* (Current Ethico-Deontological Problems of Modern Medicine), Akademiya Meditsinskikh Nauk, Moscow, 1983.

[2] Van R. Potter, *Bioethics: Bridge to the Future,* Prentice-Hall, Englewood Cliffs, N. J., 1971.

the ethics of clinical experimentation, particularly that on children. R. M. Hare (Great Britain) stated in his report that the problem in this regard was due to the fact that while medical research in child diseases could not advance without experimentation, these experiments involved physical and other types of intervention that violate the rights of children. He argued that this called for combining utilitarianism with deontology. In his comments on that report, K. E. Tranøy (Norway) generally agreed with its provisions and distinguished between two basic fields of medical ethics, namely, the ethics of clinical medical practice, and the ethics of biomedical research. Modern clinical medicine could be characterized, in his view, as a practice that presupposes a scientific theory. The ethics of biomedical research presupposes an ethics of clinical medical practice which is rooted in our ordinary morals.

Tranøy agreed with Hare in that it was very important to distinguish between deontological and utilitarian moral arguments in medical ethics, particularly when discussing experimentation on children. Deontology is concerned with norms, rules, laws, and duties, while consequentialists argue in terms of values, judgements of good and evil, etc. Tranøy considered this to be an impossible choice, for it is difficult to imagine a medical ethics that would not require equally both types of ethics.

We will later have another occasion to return to this question. For the present, let us note the importance of the ethical-humanistic approach to experimentation on man, and, more generally, to the cognition of life and man. It is important to recognize absolute values that could at least serve as "points of reference" in ethico-humanistic evaluations and decisions imposed by concrete research situations. This has been well expressed by a great humanist, Albert Schweitzer: "Only an absolute and universal purposefulness of the preservation and development of life, at which the ethics of veneration for life is aimed, is ethical. Any other necessity or purposefulness is not ethical, but is a more or less purposeful purposefulness. In a conflict between preserving my own life and destroying the lives of others or causing them harm I can never combine what is ethical and what is necessary into something that is relatively ethical, but I must choose between what is ethical and what is necessary, and in the event that I intend to choose the latter, I must be aware that I stand to be blamed for causing harm to another life. Similarly, I must not assume that in a conflict between personal and suprapersonal responsibility I can compensate for what is ethical and what is expedient by something that is relatively ethical, or more generally suppress what is ethical by what is expedient—I can only choose between the components of that alternative. If under the pressure of a suprapersonal responsibility I opt for what is expedient then I will be guilty of violating the morality of veneration for life."[1]

Schweitzer's basic idea, it seems to me, is quite productive, and could be used effectively not only in socio-ethical and humanistic evaluations of experimentation on man but also in the wider context of biomedical studies of humans.

2. A Scientific Search for Man's New Potentials vs. Neo-Eugenics; New Medicine and Bionic Technologies; the Development of Psycho-physiological Abilities and the Designing of "Artificial Intellect"; the Problem of the Unconscious and Parapsychology: Myths vs. Reality

When we turn to problems associated with discovering man's new potentials we are primarily concerned not only with using the possibilities offered by science and technology, for example, those of biomedicine, but with the limits to such activities established by society that accords priority to the preservation of human individuality, freedom, and human rights. This concerns especially manipulations of the brain and the human psyche, of human consciousness and behaviour that may produce an "identity crisis", that is a loss of a person's awareness of his place in society and of the inherent value of his personality.

Scientific and technological possibilities that neglect human realities and the social conditions of their existence, as well as the moral-humanistic aspects of the existence and development of man and the human race, generate scientistic projects that are acquiring an extremely dangerous and global character.

Let us turn to an examination (that will, of course, be very

---

[1] Albert Schweitzer, *Kultur und Ethik. Kulturphilosophie,* Zweiter Teil, C. H. Beck'sche Verlagsbuchhandlung, Munich, 1926, pp. 256-257.

fragmentary) of certain "mysteries" and puzzles, or as Darwin used to say, challenges to science in studying man both today and in the future. We will also see how all kinds of myths have emerged that capitalize on the unresolved or insufficiently resolved problems.

One of these problems today concerns man's adaptation to extremely rapid changes in the conditions of his natural-biological, social and spiritual existence. This has produced many myths concerning the future that are quite distant from reality but represent the future as something that cannot yet be verified. Let us consider how this is taking place.

Science predicts that in the future man will unquestionably greatly expand his adaptive potential with the help of a diversity of means, including pharmacology and psychotherapy, and will thus be able to function normally in the most complex extreme conditions without damage to his health. Already today major findings have been made that point to new, previously unsuspected potentials of man's biological nature and to his psychophysiological possibilities. Man's "biological equipment" reveals its universality. As "Nature's masterpiece", man must develop his organism's reserves guiding them towards harmonization with social conditions and with psychological and moral forces that he has not yet learned to maintain effectively in a homeostatic state (and which therefore produce so-called stresses, as well as all kinds of psychological anomalies that develop under conditions of emotional tension).

Yet science still faces the task of explaining many features of man's biological nature that remain a mystery to us. The greatest mystery is the human brain, the psyche as a complex of consciousness and instincts, and human intellect. Modern neurophysiology is in a process of intense growth that, some scientists believe, will produce during the next half century solutions to the major questions raised in studies of the brain during the past several centuries. This refers to studies of the brain's activity not only at the cellular and molecular level, but also at an integral level requiring a study of systemic linkages and interactions. The human brain's universal organization, which, already at the molecular level, underlies a specific capacity of neural cells for systemic functioning, sustains man's higher nervous system and the processes relating to his psyche, intellect and consciousness. To a substantial extent it is determined by hereditary factors; hence the problem of deciphering the neurophysiological code of psychological phenomena, whose solution will help achieve a more effective utilization of the human brain's potential.

This raises many complex scientific and socio-ethical problems that lead many scientists who accept the view that efforts to influence the functioning of the brain (through chemical stimulants, electrical impulses, etc.) are permissible to express serious apprehensions with regard to possible negative consequences. In particular, it is Pyotr Anokhin's view that "should attempts be undertaken at some point to make intellectual abilities the product of chemical and training laboratories, then it is entirely possible that at a subsequent higher stage of scientific development we will learn that we have produced irreversible changes in man's brain that unfortunately it will no longer be possible to remove."[1]

The challenge is to enhance intellectual activity by using more effectively the brain's existing resources which for the time being remain hidden. Yet while man's future is thus linked to perfecting functions of the brain, science has still not answered the question whether new capacities of the brain may emerge in the course of a natural evolution of some of its parts. An answer to this and other questions will make possible more accurate forecasts of man's future in connection with the development of his psychophysiological, intellectual and emotional abilities and potentials. Today it is difficult to state exactly in what direction, and especially through what methods this development will take place. It would appear that it will involve not a single method—whether natural or artificial—but rather some integrated method. We see evidence of this in the combining of neurophysiological research with engineering and even behavioural sciences. The results have already been apparent in recent decades in the achievements of biocybernetics and ergonomics, "artificial intellect" programmes and the development of psycho-

---

[1] P. K. Anokhin, "Izucheniye deyatelnosti mozga i budushcheye cheloveka" (Studies of the Brain and Man's Future), in: *Nauchno-tekhnicheskaya revolyutsiya i chelovek* (The Scientific and Technological Revolution and Man), Moscow, 1977, p. 155.

logy. Scientists see new possibilities and new reserves for man's development in each of these new approaches.

The cybernetic modeling of man's neurophysiological mechanisms and their technical reproduction in computers with a view to creating an artificial intellect that would perform an unlimited range of the natural intellect's functions opens new possibilities for the mind's "feedback influences" on brain physiology. True, many problems in this area remain unclear or controversial, including the very definition of intellect and of reason. In particular, some researchers deny altogether the possibility of creating an "artificial intellect", not to mention a mind, while others, on the contrary, see unlimited horizons.

One thing is clear: biocybernetic research will contribute to a substantial widening of man's intellectual and psychophysiological possibilities and will meet with even wider applications in ergonomics, a science that engages in integrated studies of human activities involving the use of technical equipment.

The analysis of psychophysiological phenomena under neuroemotional pressure, in particular of those involved in operating equipment, reveals not only considerable human capacities that have not been sufficiently studied but also substantial individual psychological differences among persons. Recognizing these differences and studying general psychophysiological phenomena accompanying man's activities offer one of the major approaches not only to psychophysiological adaptation but also to the development of corresponding abilities. In particular, by using the data and methods of biology, physiology, psychology as well as of engineering and socio-economic sciences, and in turning to man as a subject of activity and above all of labour, ergonomics identifies human reserves and creative potentials, including higher manifestations of the human psyche. It thus plays a role that is profoundly humane, for, as José M. Delgado rightly observed, "the direction of the colossal forces discovered by man requires the development of mental qualities able to apply intelligence not only to the domination of nature but also to the civilization of the human psyche"[1].

In particular, an even greater role will be played by psycho-

[1] José M. R. Delgado, *Physical Control of the Mind. Toward a Psychocivilized Society,* Harper & Row Publishers, New York, 1969, p. 230.

logical research based on the study not only of the neurophysiological prerequisites of the psyche, but especially of the broader interrelationship of its social mechanisms in the context of human activity. At a certain stage of the "age of biology" psychology will acquire a predominant significance, for it is particularly effective in making concrete and explaining the complex of sciences that study man, the natural-biological foundations of his existence and the higher manifestations of his creativity. The supposition has even been made that in view of the great diversity inherent in human nature, by studying the interdependence and mutual interaction of "natural man" and "cultural man", psychology will eventually be transformed into a new type of anthropology. That anthropology will embrace the problem's philosophical and political context and its value-based aspects, and will also link theoretical and experimental research to practical spheres of activity. While only time will show whether this forecast is correct, the integrated studies of man's psyche that are already taking place today provide serious grounds for forecasts of that type.

The question that arises in this connection is, would *homo sapiens* come to be replaced by some kind of "superman"? Would some new forms of human existence emerge, combined with biocybernetic devices—some "biocyborgs", as it were? Would the human race enter a new stage in its evolution at which in a substantial measure man will be created artificially—"manufactured", as it were, with the help of genetic engineering and biocybernetics which would produce a "superman" possessing extra-sensory and extra-intellectual properties—a *homo sapientissimus*?

These questions are not far-fetched; not only authors of science fiction but serious scientists as well are turning to such hypotheses and projects.

The idea of an artificially developed man, i. e., of an "engineered man", as it were, that would make him equal at least to God's antipode, a "demon", if not to God himself, has probably been around for as long as man has existed, with his propensity for fantasies, myths, daydreaming—and for scientific forecasting. It emerged initially as a dim anticipation, still mixed with myth, of the emerging force and power of scientific thinking, as a fantasy probing the future and as a fear of the coming "demons" of science. Let us recall Goethe and his *Faust,* in which he created a new "hero", Homunculus, a laboratory "twin" of man. True, Homunculus was already aware that he still needed to "dot the 'i's", i.e., to "complete" himself, so as to fully become man. Goethe observed that such beings as Homunculus were still not saddened or constrained by being a completed embodiment of man.

But already in Mary Shelley's novel, *Frankenstein. Or the Modern Prometheus,* the creation of a man-like monster leads to tragic consequences that are similar to those forecast today in many utopian projects for creating a "new man" with the help of science. Aldous Huxley, in his much talked-of *Brave New World,* carried that idea to the point of absurdity; still, it continues to trouble the minds not only of science fiction writers but also of scientists who turn, among other things, to eugenics and its new versions associated with attempts to use genetic engineering methods in order to "design" man, cloning and other methods.

The term "eugenics" (from the Greek *eugenes,* well-born) was introduced by Francis Galton in 1869, in *Hereditary Genius. An Inquiry into Its Laws and Consequences.* The book demonstrated that man's heredity, like that of any other living creature, is governed by common laws of genetics. The author also defined the problem of improving human heredity by increasing and selecting useful qualities and reducing or overcoming harmful ones. Since the latter's frequency correlates with the frequency of marriages between close relatives, efforts to limit such marriages, biochemical consulting and other measures help reduce the incidence of harmful forms of heredity. That is the aim of "negative" eugenics, which largely coincides with modern medical genetics. "Positive" eugenics, on the other hand, aims at a broader objective, namely, developing a "new man" through a selection of genotypes among the progeny of persons possessing exceptional mental or physical qualities. That branch of eugenics has been used (sometimes in contradiction with the humane intentions of its advocates) by various reactionaries and racists, particularly by the theoreticians and practitioners of the fascist "racial hygiene" and of genocide.

Naturally, the disrepute that this brought to the ideas of eugenics resulted in its failure, even though in many respects it relied on genetically well-established hypotheses and the authority of prominent scientists known for their humanistic views (Nikolai Koltsov, Hermann Muller, John Haldane, Julian Huxley and others). It is true that the basic postulate of "classical" eugenics—that concerning the possibility of guided development, through selection, of persons with exceptional intellectual abilities—was disproved by modern genetics and by the science of man, his individual and historical development. This caused many advocates of eugenics to reconsider some of its original dogmas and to turn, in particular, to social factors in human development. Some of its propositions, however, remain unchanged.

The evolution of eugenic thought in the USSR is of some interest in this regard. Often distorted, it merits at least a brief consideration. The eugenic movement in Russia, which arose long ago, was formally organized in 1920 when the Russian Eugenic Society was established in Moscow under the chairmanship of Koltsov, who also took an active part in setting up the *Russkii evgenicheskii zhurnal* (Russian Eugenic Journal) (seven volumes were published). A second centre appeared in 1921 in Petrograd where Yuri Filipchenko organized a Bureau on Eugenics that subsequently served as a basis for establishing a branch of the Russian Eugenic Society. The Bureau published *Izvestiya byuro po evgenike pri Rossiiskoi Akademii Nauk* (Bulletin of the Bureau on Eugenics of the Russian Academy of Sciences). Of particular interest for the present study are the first three issues (1922, 1924, 1925). Finally, still another institution that concerned itself with eugenics was the Medical-Biological Institute (since 1935 the Medical-Genetic Institute) under the leadership of Solomon Levit. This may be viewed as a third centre, for its studies of human genetics and medical genetics were also component elements of eugenics. Each of these centres was concerned with specific problems. In particular, the team led by Koltsov studied human heredity in pathological as well as in normal states. Its original programme for the genetic analysis of man's psychology continues to be of great interest today.[1] Geneticists led by Filipchenko were primarily con-

[1] See: N. K. Koltsov, "Geneticheskii analiz psikhicheskikh osobennostei cheloveka" (A Genetic Analysis of Man's Psychological Features), in *Russkii evgenicheskii zhurnal,* Vol. 1, Nos. 3-4, 1923.

cerned with analyzing the genetics of gifted persons through survey questionnaires addressed to scientists, representatives of the arts, and students, as well as with developing a system of eugenic consultations for young persons considering marriage.[1] The programme of the third centre envisaged further advances in the twins method for in-depth studies of the role of mechanisms of interaction between genetic and environmental (social) factors in man's ontogenesis, studies of the genetics of a number of diseases (diabetes, daltonism, allergies, hypertension, ulcers), and analyses of progeny from marriages among closely related persons.[2] Foreign anti-fascist geneticists, including John Haldane, L. Hogben, Muller, and G. Dalberg also contributed to the work of the Institute.

In an overall evaluation of the eugenic movement in the USSR let us note a significant point: all the programmes mentioned were concerned mainly with solving strictly scientific problems. The situation was described by Koltsov as follows: "For the time being the eugenic movement is still very young and is concerned mainly with studying human heredity and mutability. But it has set before itself a noble goal—to perfect human nature. Its attainment will be science's greatest miracle. But the paths leading to its practical achievement are difficult, the work will require centuries, and one must prepare well ahead of time with the conscious participation of the entire population."[3]

A democratic orientation and a responsible attitude towards using the laws of human heredity and mutability in order to protect and improve the health of future generations are important traits of Russian eugenic thought. One may see from the programme of the Bureau on Eugenics published by Filipchenko

[1] The programme of the Bureau on Eugenics is presented in the popular pamphlet *Chto takoye evgenika?* (What is Eugenics?) by Yu. A. Filipchenko (Petrograd, 1921).

[2] The substance of that programme was published in *Medikobiologicheskii zhurnal* (Medico-Biological Journal), No. 5, 1929, in articles by S. G. Levit and A. S. Serebrovsky. See S. G. Levit, "Genetika i patologia (v svyazi s sovremennym krizisom meditsiny)" (Genetics and Pathology and the Modern Crisis in Medicine); and A. S. Serebrovsky, "Antropogenetika i evgenika v sotsialisticheskom obshchestve" (Anthropogenetics and Eugenics in a Socialist Society).

[3] N. K. Koltsov, *Chudesnye dostizheniya nauki* (Remarkable Achievements of Science), Moscow, 1927, p. 28.

the great importance that was attached to campaigns of popular education in genetic (eugenic) knowledge among wide masses of the population. In particular, in programme stated: "If ... eugenic ideas will be disseminated and will become widely accessible in schools, where everyone's world view is formed, then attitudes towards these questions will change. One may then hope that mankind will consciously and voluntarily develop a number of mandatory norms in that regard without using any harsh measures. Once again, what is important is knowledge, and any knowledge inevitably yields results. Accordingly, in the field of eugenics the entire problem is to produce such knowledge and disseminate it."[1]

One can easily see that these views are in many ways consonant with those of Galton, the founder of eugenics. In particular, Galton assumed that the eugenic movement had to pass through three stages. At first it would be concerned simply with the cognition of the laws of human heredity and mutability and with a theoretical elaboration of a set of measures aimed at improving the human race. The second stage would be of a practical character; at that point, eugenic measures would acquire the status of laws. After persons had become conscious of the need to be guided by eugenic rules in everyday life and the need for laws would no longer exist, a third stage would begin in which eugenic enthusiasm would become mankind's Bible.

Galton's idea concerning a "eugenic religion" was endorsed by Koltsov, who wrote: "This description of eugenics is that of a religion. Cultivated mankind has always been inspired by religion—an ideal that is perhaps distant and unclear but that has guided men in their lives and in solving problems of good and evil... Eugenics, too, has defined a lofty ideal that is also worthy of dedicating lives to and would motivate men to sacrifices and self-imposed constraints for the sake of creating, through the conscious work of many generations, a higher type of man who would rule in Nature and be a creator of life... Eugenics is the religion of the future now awaiting for its prophets."[2]

---

[1] Yu. A. Filipchenko, *Chto takoye evgenika?*, p. 27.

[2] N. K. Koltsov, "Uluchsheniye chelovecheskoi porody" (Improving the Human Species) in: *Russkii evgenicheskii zhurnal*, Vol. 1, No. 1, 1922, p. 273.

It should be noted that eugenic thought that was thus emerging in Russia was strongly influenced by Galton's ideas. Above all this concerns the cardinal questions of eugenics, namely, its ideal and ways for achieving it; attitudes with regard to "negative" and "positive" eugenics; the problem of inheriting man's psychological and intellectual qualities; the problem of degeneration, as well as the issue of eugenics' social status. Subsequently, however, eugenic thought in the USSR progressed rapidly, and its development was not without its complexities and contradictions. These are attributable to a number of scientific, ideological, and practical reasons.

To a large extent the changes in the views of eugenicists reflected new knowledge in human genetics. In particular, Sergei Chetverikov's elaboration of the principles of popular genetics and their use in analyzing human populations made it possible to establish on scientific grounds the ineffectiveness of "negative" eugenics (in particular of sterilization) and to raise doubts concerning the possibility of developing a eugenic ideal. A departure from narrow autogenetic views and a recognition of the substantial role of environmental factors in manifestations of inherited attributes made it possible to show that efforts to combat hereditary diseases could be undertaken on an ontogenetic level by creating conditions under which the effectiveness of defective genes would be minimal. The field of medical genetics developed intensively. With the help of improved methods of chromosome research and cultivation of leucocytes Soviet geneticists laid the foundations for analyzing man's chromosome complex and came closer to an understanding of the latter's significance for diagnosing a wide range of hereditary diseases caused by chromosome anomalies.

Successful joint research by geneticists, psychologists and physicians established that man differs from animals fundamentally in inheriting specific psychological and intellectual qualities whose development greatly depends on social factors. At the same time, the cooperation within the eugenic movement of scientists from various specialized fields contributed to a better understanding of the complex structure of eugenics and of its inherent differentiation into biological, medical and social disciplines, whose own further development has almost completely absorbed the subject of eugenics proper.

Scientific progress and the new socio-economic conditions that emerged following the October Revolution of 1917 in Russia contributed to the disappearance of eugenics as a general scientific approach to man and to his future. It was replaced, on a strictly scientific basis, by the genetics of man, anthropology, medical genetics, and, most importantly, by the social sciences and social practice. A particularly strong blow to eugenics was dealt by the Marxist-Leninist conception of forming a new man, an approach based on dialectical and historical materalism, the theory of scientific communism, and real, communist humanism and morality. This was entirely natural, for the normative aspect of eugenics and its social component had rested on a utopian conception of social phenomena, a biologistic interpretation of man's essence and an abstract humanistic ideal of his future.

This naturally influenced the evolving views of eugenicists in this country, particularly those of Koltsov and Serebrovsky, whose approach to the social positions and conclusions of "classical" eugenics was largely critical.

Koltsov substantially revised his earlier position concerning the possibility of evaluating human qualities from a eugenic point of view. In an article entitled *"Geneticheskii analiz psikhicheskikh sposobnostei cheloveka"* (A Genetic Analysis of Man's Psychological Potential) he emphasized that is was precisely the human race's enormous diversity of genotypes that was its greatest and most valuable specific attribute which ensured man's progressive evolution in the presence of the most diverse uncertainties concerning its future.[1] That standpoint differs substantially from the view that he had expressed earlier that human populations possess some kind of a eugenically valuable group. The new standpoint was expressed most clearly in an article by Koltsov entitled *"Rodoslovnye nashikh vydvizhentsev"* (Our Outstanding Personalities' Family Tree) on the genealogy of such persons as Maxim Gorky, Leonid Leonov, Fyodor Chaliapin and Sergei Yesenin.[2]

Thus, as P. Rokitsky correctly notes, "it may be said that

---

[1] See: *Russkii evgenicheskii zhurnal (Russian Eugenic Journal),* Vol. 1, Nos. 3-4, 1924, p. 307.

[2] *Russkii evgenicheskii zhurnal (Russian Eugenic Journal),* Vol. 4. Nos. 3-4, 1926.

Koltsov was a 'moderate' eugenicist, who interpreted correctly the possibilities for overcoming genotypical shortcomings with the help of appropriate education and care. He did not share the extreme views of some advocates of eugenics who proposed to use the sperm of individuals considered to be eugenically most valuable for artificial insemination. Such ideas were alien to the spirit of humanism that characterized Koltsov's work."[1] At the same time, he retained many erroneous views which followed from the excessive importance that he attached to the role of biological factors in human development.

The position held by Serebrovsky concerning extreme views in eugenics was in many respects similar. In particular, he ridiculed Galton's opinion that a child born into a family of the privileged classes could sometimes be valued a thousand times higher than that born into a worker's family. To begin with, Serebrovsky questioned the scientific grounds of that conclusion. "We must state emphatically," he wrote, "that our knowledge of human genetics and of the transmission of individual valuable human properties through heredity is still . . . so fragmentary that we should never pretend that our conversations to the effect that particular classes are better than others in terms of their hereditary properties are scientific conclusions."[2] He stressed that while there were indications that talents were spread unevenly throughout the human race, this was by no means sufficient for a eugenic evaluation of such large social groups as classes. Accordingly, "statements to the effect that the group of the intelligentsia or the group of the bourgeoisie constitutes the salt of the earth while all the rest constitute soil or muck, as it were, needed for the salt of the earth to flourish, are so ludicrous that only their class blindness can prevent those making this claim from seeing this."[3] While rejecting reactionary eugenic ideas he stressed the real problems of human evolution: what is happening to man's biological nature, what genetic processes are taking place in human populations in the absence of natural selection, what are the consequences of mutagenesis for the human race's gene pool?

---

[1] P. F. Rokitsky "Nauchnye vozzreniya N. K. Koltsova" (The Scientific Views of N. K. Koltsov), in: *Voprosy filosofii,* No. 7, 1972, p. 99.

[2] *Pod znamenem marksizma,* No. 3, 1926, p. 113.

[3] *Ibid.,* p. 114.

Nevertheless, while rejecting reactionary bourgeois conclusions based on eugenic theories, Serebrovsky did not always succeed in avoiding imprecise formulations and errors. While stressing the importance of a biological approach to the human race's future, he did often refer to social factors. For example, he observed that if it is indeed true that artistic and scientific creativity and organized human activities to a certain extent "rest on a hereditary basis, then we are justified in viewing the set of genes that produce exceptionally talented persons in human societies, or conversely, idiots, as a form of national wealth, a gene pool from which the country draws its people."[1]

As for the relative value of different population groups, Serebrovsky viewed that issue with considerable caution: "I wish to repeat that there our knowledge is still so limited that the very posing of such questions is completely premature."[2] In that connection he sharply criticized the numerous eugenic programmes proposed in capitalist countries.[3] At the same time, Serebrovsky shared the erroneous view that it is possible to apply to human beings some of the techniques used to improve the heredity of animals (for instance, artificial insemination). His approach to this question was based on unfounded presumptions concerning the withering away of the family and the disappearance of marriage. Eventually, however, criticism of that view and the socio-political conditions in which he lived and worked helped him recognize this error as well.[4]

---

[1] *Pod znamenem marksizma, op. cit.*, p. 116.

[2] *Ibid.*, p. 115.

[3] See: A. S. Serebrovsky, "Antropogenetika i evgenika v sotsialisticheskom obshchestve" (Anthropogenetics and Eugenics in Socialist Society), in: *Trudy kabineta nasledstvennosti i konstitutsii cheloveka pri Mediko-biologicheskom institute* (Transactions of the Office of Heredity and the Human Constitution of the Medico-Biological Institute), Issue No. 1, Moscow, 1929.

[4] It is true that at times that criticism became less than competent and had negative consequences, as noted by D. K. Belyayev and P. F. Rokitsky: "Unfortunately, sharp and occasionally incompetent criticism of eugenics and the noisy campaign that developed around this issue and extended beyond the framework of science was also detrimental to work on the problem of human heredity as a whole, resulting in a complete cessation of research in human genetics in the USSR for a relatively long period of time." (D. K. Belyayev, P. F. Rokitsky, "O nekotorykh

In a general evaluation of the views held by the leaders of the eugenic movement in the USSR, one must agree with Nikolai Bochkov's conclusions: "Their works do contain certain elements of simplification and even contradictions, especially the works of N. Koltsov. Nevertheless, when viewed against the background of similar research in other countries, at the peak of eugenics' achievements our scientists' approaches were marked by a progressive orientation, a more rigorous scientific character and a democratic interpretation of their findings. For example, Koltsov and Filipchenko correctly addressed the problem concerning the role of the social environment in the realization of individual abilities. They pointed to the presence in the people of an immense fund of inherited gifts that did not find expression in pre-revolutionary times because of the absence of corresponding conditions. They rejected completely approaches to improving human nature that are based on coercion."[1]

Similar views were advocated at about the same time by Hermann Muller, who worked in the USSR for several years and who actively campaigned for eugenic ideas. This prominent scientist who made a substantial contribution to the development of genetics, announced a "crusade" on behalf of eugenic measures that were destined, in his opinion, to save the human race from a genetic catastrophe. Initially he proposed the introduction of controls over human reproduction and of selection for improving the human gene pool, to be followed by a programme of artificial insemination of women using preserved sperm from outstanding men, drawn from specially created sperm banks. This idea was discussed in Muller's book *Out of the Night: A Biologist's View of the Future.*[2] At the Second International Congress on Human Genetics (1961), he elaborated that idea in his conception of embryo selection; he noted, however, that he could not guarantee a complete success since the manifestation of human attributes also depends on the social environment and

---

metodologicheskikh problemakh v razvitii sovetskoi genetiki" (Concerning Certain Methodological Problems in the Development of Soviet Genetics), in: *Voprosy filosofii,* No. 3, 1977, p. 143.

[1] N.P. Bochkov, *Genetika cheloveka* (Human Genetics), Meditsina Publishers, Moscow, 1978, p. 18.

[2] H. J. Muller, *Out of the Night,* Vanguard Press, New York, 1935.

development.[1] "Only the impending revolution in our economic system," Muller stated, "will bring us into a position where we can properly judge, from a truly social point of view, what characteristics are most worthy of a man...

"Thus ... it is up to us, if we want eugenics that functions, to work for it in the only way now practicable, by first turning our hand to help throw over the incubus of the old, outworn society."[2]

As we see, Muller associated the application of eugenic ideas with specific socio-economic changes in capitalist society. But even the purely scientific aspect of these ideas has met with objections from a number of geneticists, including Theodosius Dobzhansky, George Beadle and Bentley Glass. Still, eugenics has many proponents among modern scientists, some of whom, moreover, have moved beyond theoretical support to practical activities.

Today such projects are often associated with applying to man the methods of genetic engineering, cloning, etc.[3] At the same time, even when neo-eugenic ideas are accepted in principle they often find new substantiations and new forms of embodiment. Interesting data and conclusions concerning the danger of the human race's "genetic degradation" and the ensuing need to adopt eugenic measures are cited, for example, in the survey entitled *Biology and the Future of Man.*[4] Although its authors do not define concrete social conditions for applying their humanistic position, they, nevertheless, reject the irresponsible manipulatory ideas that are being revived today within the framework of so-called neo-eugenics.

A number of British scientists have also addressed the need

---

[1] Hermann J. Muller, "Means and Aims in Human Genetic-Betterment", in: *Ethical Issues in Modern Medicine*. Ed. by R. Hunt and J. Arras, Mayfield Publishing Company, Palo Alto, California, 1977, pp. 60-64.

[2] Quoted in: Martin P. Golding, "Ethical Issues in Biological Engineering", in: *Ethical Issues in Modern Medicine*, p. 72.

[3] See, for example: *The Cloning of Man. A Brave New Hope— or Horror?*, Ed. by Martin Ebon, A Signet Book, New York, 1978.

[4] *Biology and the Future of Man*. Ed. by Philip Handler, Oxford University Press, New York, 1970. For further details, see: I. Frolov, *Progress nauki i budushcheye cheloveka* (Scientific Progress and the Future of Man), Politizdat, Moscow, 1975.

for giving greater weight to social conditions in discussing potential projects associated with a reconstruction of human genetics.[1] They even state that it is desirable to prohibit certain types of research in that sphere in view of the possibility that the latter may be used to the detriment of the human race. Many consider that such knowledge is in itself so dangerous that its dissemination should be limited. In a civilization in which the individual is sacred, they state, eugenic procedures are unethical.

The advocates of the most recent trends in eugenics, including Elof Carlson, Bernard Davis, Joshua Lederberg and others, seek to apply the latest advances in modern genetics. In other instances, the "spectre of eugenics" is associated not so much with contemporary genetic research, as with ... ideas of revolution and of creating a new man.[2]

To a substantial extent, neo-eugenics, like eugenic projects in general, seeks to find scientific and personal-emotional support in the idea of an "all-absorbing" concern for man and the human race, for their dignity and freedom, and for their future. In this context it is assumed that a person who has been subjected to eugenic measures (those of "positive" eugenics) will correspond more closely to his essence. In terms of Muller's classification, such a person would possess a stronger health physically, have stronger mental powers and creativity; morally he would be warmer and more sincere in his sympathy to others, and have collectivist inclinations; his perceptions would yield a richer understanding of reality and find better forms of expression.

Characteristically, compared to old eugenics, neo-eugenics places a greater emphasis on the means to be used in implementing its projects, on their morality, and on their moral admissibility. Generally, what is proposed is "a noble human form of eugenism" (P. Teilhard de Chardin) that must be applied gradually in the course of centuries, and on a voluntary basis. Moreover, the positions of extreme scientism and social-biologism, defended by a number of modern neo-eugenists, are being subjected to fundamental criticism "from within", as it were. Obviously, this criticism, motivated by considerations of human-

---

[1] *The Social Impact of Modern Biology,* Routledge & Paul, London, 1971.

[2] Charles Frankel, "The Specter of Eugenics", in: *Commentary,* Vol. 57, No. 3, March 1974, pp. 25-33.

ism and an awareness of the social responsibility of scientists, plays an extremely important role in opposing neo-eugenics and is characteristic of many scientists today.

In short, among Western scientists as well, aside from advocates of neo-eugenics and of other approaches that allegedly open the way to a new "fabricated superman", there are many who are guided by a realistic and sober approach to reality and who are proposing to consider not only the available scientific and technological possibilities, but also the no less important socio-ethical aspects of the problem. True, certain advocates of eugenics likewise seek to apply the methodology of a "balanced" social-biologism that would combine eugenic projects for creating the "ideal man" with social ones. This was clearly seen in the works of Koltsov, Serebrovsky and also Muller, who believed that programmes of planned eugenics would provide the opportunity to make "unlimited progress in the genetic constitution of man, to match and re-enforce his cultural progress and, reciprocally, to be re-enforced by it, in a perhaps never-ending succession."[1] In some respects Muller's position in this regard is close to that of a number of scientists who continue to advocate the idea of creating a "socialist eugenics".

Marxists, however, cannot accept the ideas of neo-eugenics even in an "ennobled" form, and not only because these ideas have compromised themselves in the past. Marxist-Leninist theory concerning man and the ways of his development needs no such "additions", for it already incorporates the findings of science, including genetics, made in the study of man. This theory addresses the problem of moulding the new man as social in essence, and it is only from such a position that it views biology above all genetics, that is increasingly serving the needs of man, his health and development.

Neo-eugenic projects for creating the "ideal man" are unsound, above all, in terms of scientific theory, for they rest on a knowledge of human genetics that is still extremely limited and on false conceptions that link man's genetic foundations directly to his mental and, more generally, spiritual qualities. These projects are fallacious from a social standpoint, for they rely ex-

---

[1] *The Guidance of Human Evolution in H. Muller's "Studies in Genetics"*, New York, 1962, p. 590.

clusively on genetic factors, rather than on social ones, as regards changing man. Not surprisingly they have been adopted by advocates of racist ideologies, and of the theory and practice of genocide. Neo-eugenic projects are also misleading in terms of their philosophical, world-view and methodological aspects, for they present a distorted view of man's essence and of his role in the world and in history. They also stand to be condemned from a humanistic point of view because they infringe upon the inviolability and uniqueness of the human personality by promoting scientistic and manipulative approaches to the latter. Neo-eugenic projects should also be rejected for moral and ethical reasons, as their implementation threatens the human race and raises doubts concerning such basic values of human existence as love and parental sentiments.

This does not mean, of course, that any intervention into man's heredity is impracticable and undesirable as a matter of principle, and that there will never be any real prospect for changing man's biological nature in a desired direction even in the distant future. One must, however, distinguish between scientific feasibility and actual practice. The latter cannot be guided by abstract presumptions and requires a concrete definition of the social conditions that must be met if particular ideas are to be applied. At present, neo-eugenic projects can objectively play and are in fact playing a reactionary social role. I am deeply persuaded that their application would produce a genetic catastrophe for the human race that would be far more dangerous than the one from which eugenics promises to save us.

While rejecting neo-eugenics for purely scientific as well as social, philosophical, humanistic and ethical reasons, we cannot ignore the real prospects for man's biological development that are emerging from the research in human genetics that has developed intensively in recent years, particularly in the field of genetic engineering. That research, which has nothing in common with neo-eugenics, reveals new and occasionally equally complex problems. It is therefore evident how important it is to discern the social aspects of genetic-anthropological research and their humanistic orientation, which precludes scientistic and manipulative approaches to man and is based on respect for man's freedom and uniqueness.

In a "divided world" any possibility of influencing man so as to produce profound changes in him is very dangerous; if used, it can only bring new hardships. Moreover, science itself is somewhere at the very beginning of the process of man's cognition, and what we do know is far less than what we still do not know. At such a time not only bad intentions but also "pseudo-scientific ignorance" in the form of scientistic self-assurance can cause irreparable damage and is therefore extremely dangerous.

As for the very distant future one should expect, I think, some major developments as science will usher in the "age of man", and the full power of scientific cognition will focus on man. But this requires corresponding social conditions based on reason and humanity that will also produce a new scientific ethos. At that stage an awareness of the uniqueness of *homo sapiens et humanus* will develop, and it is not for us to anticipate the conclusions that will follow. In any event, it will become possible to carry out successfully projects for changing man's biological nature only at the concluding stage of the "age of biology", after social homogeneity of the human race will have been achieved. This factor will eventually make it possible to perfect man biologically in accordance with the ideal that was created in myths and utopias.

The above approach, which recognizes the problem's diverse humanistic dimensions, should apparently be extended not only to neo-eugenics but also to projects for creating *homo sapientissimus,* "the extremely intelligent man", or *machina sapiens,* "biocyborgs", through genetic "designing".

Acute socio-ethical and humanistic problems arise in connection with mind-altering influences. The appalling consequences of such manipulations have made psychosurgery and other artificial methods for affecting the personality an instrument of violence used for suppressing independent thinking and protests.

Obviously, the above does not imply a complete rejection of the idea of influences that alter the brain and psyche of mental patients in ways favourable for the latter. Still, with today's bitter struggle among the world's different political forces it is hardly practicable to make the above a primary goal, as its attainment will inevitably become just another treacherous and repulsive instrument in that struggle.

With regard to the question of psychosurgery's legitimacy the previously mentioned WHO document entitled "Protection of

Human Rights. In the Light of Scientific and Technological Progress in Biology and Medicine" stressed that psychosurgery represents one of the most difficult bioethical problems. Besides, the scientific potential and consequences of psychosurgical intervention being still insufficiently clear, in the opinion of a number of its proponents such intervention is permissible only in the case of exceptional chronic diseases, if at all. The document notes that many opponents of psychosurgery regard surgical manipulations on the brain in cases of purely behavioural aberrations of unknown etiology as inadmissible under any circumstances. In that connection the conclusion was that since there are wide differences concerning the legitimacy of applying psychosurgery, and in view of the uncertain outcome of such intervention as well as of the absence of established theoretical grounds, research had to be continued in order that this method of treatment may be evaluated.

Since even cases of pure therapy give rise to serious concern, the same obviously goes for plans for actively influencing the brain's activities in order to alter it and produce a "superman" —a *homo sapientissimus*—at a time when there is an obvious "lack of firm theoretical grounds" for that. Besides, as Jérôme LeJeune, a French geneticist, humorously noted, in order to create a man who is more intelligent than we are, we must already be more intelligent than we can be. In LeJeune's view, manipulations in genetic engineering will not create a "supermankind"; only a close alliance of the heart and reason will one day save the human mind.

This excellent statement reflects a truly humanistic approach to the problem. It must be fully applied to any programme seeking to achieve a psychophysiological perfecting of man and an intensification of his intellectual activities, whether with the help of natural or artificial influences on the brain and psyche, or through the development of technical means to model individual functions of intellectual activity. In particular, the above is evident in attempts to create an "artificial intellect". In the opinion of a number of scientists, existing approaches to that problem involve numerous simplifications. The scientists believe that unless the human intellect's psychological mechanisms are also recreated, what engineers will be recreating will not be human thinking but rather the embodiment of their own simplified conception of the latter. The development of biocybernetics continues to reveal new possibilities for modeling intellectual activities—and possible limitations. Man seeks to make computers and computer systems resemble as closely as possible the complexity and variety of functions of the intellect. It is clear that machines can imitate the activity of the human brain only to the extent that it can be described in algorithms and programmes, and only in corresponding areas. Yet these areas are quickly expanding, and while we have not yet encountered any rigid barriers this does not mean that none exist. Even though computers are already capable of some "inherently human" thinking and sensory operations, thus drawing closer to recreating the higher functions of the human brain, scientists note that no one has yet seriously posed the problem of creating an artificial intellect on a par with that of Dostoyevsky or Einstein. At the same time, one cannot ignore the rapid progress taking place in cybernetics and electronics. Scientists believe that there exists a "complexity threshold" based on intellect's social nature, for man is a product of not only biological but of social evolution as well. His intellect's "social component" is governed by will and motivation, as well as by the "collectivistic" nature of the cognition process—a factor that makes man a system with a practically infinite number of degrees of freedom.

Naturally the "inherently human" characteristics of brain activity are not determined by this alone, as it is hardly proper to view the extremely complex world of the psyche and of human consciousness, as well as man's moral and aesthetic emotional experiences, in isolation from other manifestations of human reason. What new possibilities are offered in this regard by the development of biocybernetics? In their forecasts, scientists not only envisage the possibility of the development of a "creative artificial intellect", but are also considering the potential dangers for man that it can pose. A more cautious view envisages instead the possibility of altering man's biological characteristics in order to enhance his intellect without any limitations through a symbiosis with the "artificial reason".

The outlook for the creation of what I describe as a biocyborg is assessed quite optimistically not only by science fiction writers but also by a number of serious scientists. In particular, according to a forecast of the RAND Corporation's research and forecasting centre prepared in 1972, a "man-machine" symbiosis designed to enhance man's mental abilities through direct interaction between the brain and computers will be demonstrated as early as in 2012. Viktor Glushkov, a Soviet scientist, predicted that by the year 2020 research scientists will have achieved a full symbiosis of man and machines. Moreover, in the opinion of a number of scientists, (John Bernal, Glenn Seaborg and others) eventually it will be possible to transmit thoughts directly from one brain to another without the intermediacy of words, and not only thoughts but also emotions and reminiscences. Such forecasts, however, go beyond the usually accepted time frame for scientific determination of human prospects.

However, that issue also involves a factor that is already today greatly relevant in studying the human intellect, psychophysiological phenomena, the mechanisms of memory and intuition, the unconscious and various forms of human sensitivity. These phenomena are interpreted in modern science in highly diverse ways that are occasionally accompanied by all kinds of myths and anti-scientific notions, as is the case in psychoanalysis and parapsychology. It would appear that science is now on the threshold of a new approach to the study of unknown forces of the human organism that calls for new research methods and a new way of thinking.

The same applies to the problem of the unconscious, a problem with ancient origins, which still has not been fully explained in scientific terms. According to Freud's theory, the psyche consists of three systems, namely, the conscious, the subconscious, and the unconscious; the latter includes instinctive notions, and supplanted experiences, expressed as spontaneous creative and destructive forces, sexuality and aggression, life and death. While the psychoanalytic conception of Freud and his followers opened the way to the treatment of many neuroses it did not offer a satisfactory explanation of the problem of the unconscious. More, in its general theoretical and philosophical conclusions, as is known, it leaves the domain of science, as do various neo-Freudian interpretations of the unconscious.

A scientific approach to the unconscious, on the other hand, makes it possible to identify the socio-personal and psychophysiological mechanisms underlying that complex phenomenon which plays an important role in personality formation and behaviour and in man's creative activities. It would appear that attempts to model and reproduce technically certain fragments of the unconscious, which enhances man's ability to comprehend the world and enjoy it, would be useful in this regard. The emphasis, however, should be placed on studies of the untapped potential of "natural" forms of life, including man's psychophysiology, the functioning of his brain, sensory mechanisms, self-regulation, and, more generally, a study of all those forces that support, in ways still largely mysterious, the functioning of the human organism, primarily mental activities.

In this connection, several observations would be in order concerning so-called parapsychology, which has, over a certain period, particularly in recent years, caused much clamour, and which is often advertised as a new approach to unexplained psychological phenomena.

Parapsychology defines as its field of study various forms of sensitivity that make it possible to receive information in cases when its reception cannot be explained in terms of the activities of known sensory organs. It also studies the influences exerted by living organisms with regard to physical phenomena taking place outside the organism without the intermediacy of muscular efforts (e. g., by a desire or a mental influence). The forms of sensitivity that parapsychologists have in mind include telepathy and clairvoyance, while the physical influences comprise psychokinesis, psychophotography and paramedicine.

While much research through the world is focused on parapsychological phenomena, and especially on extrasensory perception, the basic problem that remains is that of the authenticity of parapsychological phenomena. Leaving aside clearly fraudulent and mystical speculations of parapsychologists that result in sensational exposures, it should be recognized, specialists believe, that "some of the so-called parapsychological phenomena apparently do exist. But a recognition of their existence is made difficult by the fact that the channel through which information or influence is conveyed remains unknown. Major hopes and efforts in this regard are currently focused on studies of electromagnetic fields emanating from organisms, viewed as a means of biological communication and as information carriers. Such studies are conducted on insects, animals and man; in recent years, however, many researchers, at least outwardly, did not associate their work with parapsychology. The physical basis of these phenomena has not yet been established."[1]

This points to the conclusion that since the concept of parapsychology combines "supernatural" phenomena with phenomena that actually exist but have not yet been given a satisfactory scientific explanation, the former must be exposed for what they really are and demythified while the latter have to be studied on the basis of psychological, physiological, biophysiological, medical and other methods. It is only in this way that one can hope to establish the true nature of the phenomena described by parapsychology and that of many other unsolved mysteries of the human mind.

It is difficult to foresee what this will bring to the human race, what new forces will be brought to life and how this will influence man's life and future. Yet is such foresight really necessary at this stage, when we are still at the very beginning of the road leading to new mysterious forms of knowledge? As Norbert Wiener, "the father of cybernetics", wisely observed, the greatest error in forecasts is an excessively narrow projection into the future of science's current possibilities. We will therefore seek to avoid that error in the hope that the unraveling of new secrets of human psychology will serve man's wellbeing and will make him more powerful, more reasonable, and more humane.

The goal of science today, as I understand it, is not only to study man's unused potential, including that associated with his psychophysiological development, but also to avoid approaches in experimentation that endanger human beings. This refers both to neo-eugenic and biocybernetic utopias and to parapsychological conceptions whenever they are declared to be "inaccessible" to objective analysis and are accompanied by idealistic, mystical and sometimes blatantly fraudulent interpretations.

The greatest mystery of the biology, genetics and psychophysiology of man concerns the integrated functioning of all his internal forces, determined qualitatively by the social factors which constitute man's essence. Man's inner world is governed by many forces produced by the interaction of phenomena that

---

[1] V. P. Zinchenko, A. N. Leontiev, B. F. Lomov, A. R. Luriya, "Parapsikhologiya: fiktsii ili realnost?" (Parapsychology: Fiction or Reality?), in: *Voprosy filosofii,* No. 9, 1979, p. 135.

are familiar with others that are still unknown. At various times and in various teachings—both idealistic amd materialistic—these phenomena were given different names: "entelechy" (Aristotle, Hans Driesch, and others); "vital force" or "soul" (some vitalists, Friedrich Schelling and others); *"élan vital"* (Henri Bergson); *"holos"* (Jan Smuts); "equifinality" (Ludwig von Bertalanffy; "systemic forces" (Ervin Bauer); "integron" (François Jacob); and "biofield" (Alexander Gurvich and other contemporary interpreters of parapsychological phenomena). Leaving aside idealistic interpretations of the integral "systemic forces" of living organisms, including man and his psyche, we cannot fail to recognize that apparently it is precisely there that one should seek the key to a materialistic explanation of mental phenomena such as the unconscious that we still view as "mysterious". This, however, presupposes a new stage in the development of the sciences of man, including psychology, in which there will be a transition to truly integrated and system-oriented studies of the above phenomena.

An objective scientific analysis of interactions among the "systemic forces" will eventually make it possible to clarify many mysteries of the human psyche and to discover new biological and social human potentials.

Yet the above reflects only one aspect of the potential of scientific cognition. Another one that is equally important is a value-based humanistic aspect involving the very foundations and meaning of human existence. Some scientists predict that eventually scientific analysis will cover all of man's biological and intellectual activities, describing them in mathematical formulas. Similar forecasts have also been made concerning man's inner world, including emotions, instincts, his most complex moral experiences, etc. But what if the above is only some "regulatory objective" that may be sought by scientific cognition but is unlikely to be attained in the foreseeable future? Moreover, will cognition always pursue such an aim? Perhaps men will eventually understand that they will perish as humans unless they leave untouched something that belongs only to their uniquely human individuality and cannot—and must not—be translated into the language of mathematical formulas, no matter how enticing this may be for *homo sapientissimus,* the "fabricated man", or *machina sapiens,* a "biocyborg".

Boris Astaurov, a prominent Soviet scientist, observed prophetically that "one must desire and believe that as man will influence his environment in ways that are increasingly rational, creating for himself an increasingly perfect habitat, and as he will find increasingly humane and effective ways for perfecting his heredity, the creatures of evil and darkness will recede before the spirits of virtue and light. There is no doubt that in a society based on social justice and on lofty communist ideals the social sphere, which directly influences the contradictory nature of heredity, will promote complete development of all of man's inborn qualities that contribute to the development of humaneness and altruism, and, conversely, will suppress manifestations of aggressiveness and egoism, inherited by man from his zoological ancestors. One must hope that as a result of that process both the future environment of the entire human race and its future heredity will merge harmoniously, eventually becoming those needed to create a *homo sapiens et humanus.*

"This goal must be pursued by the highly moral persons living in progressive society, men to whom both the present and the future belong. But they must also remember that even though their own evolutionary development has entered into a social phase, the laws of biology that effect their environment and organism are still in effect. They must also keep in mind that the law of nature with which we began, namely, that each of man's characteristics depends not only on his environment but also on his heredity, and that in becoming the master of his future and in taking mankind's evolution into his hands he must learn to deal with extreme care and caution, wisely and humanely, not only with his environment but also with his heredity.

"In the social phase of man's evolution, the new name which he must earn would be *homo sapiens et humanus,* the thinking and humane man."[1]

He should subordinate everything to the attainment of this goal including science, which should also be seen in the light of reason and humaneness. This is the only prospect that is worthy of a thinking and humane man.

---

[1] B. L. Astaurov, "*Homo sapiens et humanus*—chelovek s bolshoi bukvy i evolyutsionnaya genetika chelovechnosti" (*Homo Sapiens et Humanus*—Man with a Capital "M" and the Evolutionary Genetics of Humaneness), *Novy mir,* No. 10, 1971, p. 224.

3. Humanistic Problems of Human Genetics; Genetic Engineering: Unlimited Possibilities and Possible Limitations

Today the humanistic "dimension" of science and its socio-ethical regulation in accordance with the interests and needs of man are the subject of intense discussions and of solutions in the light of the new developments associated with possible aplications of recent advances in genetics, and especially human genetics. Many current studies are concerned with analyzing the social and ethical, humanistic problems of human genetics. We will try to generalize the results of these studies and add new data and considerations which reflect the latest developments in this field and the corresponding new themes in discussions. It would appear that a review of past discussions (particularly those on the ethical problems of genetic engineering) in the light of the present situation will be useful in understanding not only the course of development of socio-ethical thought in this field, whose history is limited to a single decade, but, more generally, the specific ways in which problems of the ethics of science are being defined and solved today.

Let us recall at least some of the momentous events that have unfolded in the recent past, producing heated discussions of socio-ethical and humanistic questions. One should emphasize again that such an intense concern with the socio-ethical aspect of genetic engineering was due to the fact that it turned out to be the point of intersection of many lines of discussion and polemics developing in different directions. Aside from its purely academic interest, genetic engineering soon became a matter of quite practical interest. Today it is even viewed as a prototype of the technologies of the future (e. g., biotechnology) that will rely much more than current technologies do on advances in biological science. There is much discussion of the prospects that genetic engineering offers in medicine and agriculture. Beyond this, the development of genetic engineering has led many scientists to reflect on science's social responsibility and on the feasibility of both "internal" and "external" social regulation of scientific research. Finally, both large segments of public opinion and even government and legislative bodies have become involved in discussions of the problems of genetic engineering.

At a previous stage, the emergence and development of cybernetics attracted attention to control processes and to the fact that in the latter weak energy signals lead and control high-power energy transformations. Similar problems are addressed by genetics which is concerned with the structure and mechanisms that control the growth and development of biological organisms. Genetic engineering seeks to enable man to intervene directly in the regulation of the genetic programme of organisms by restructuring them in desired ways. As a result, very delicate micro-structural transformations of individual cells may produce substantial effects at the macro-level of organisms, populations, and even the biosphere as a whole. Thus man acquires a power that must be used with the greatest caution. It is precisely this point that ultimately defines the *socio-ethical* content of studies in genetic engineering.

The British author Aldous Huxley, who took an interest in the impact of scientific development on man, was probably the first to point to the dangerous consequences inherent in intervening in human genetics. In the foreword to the 1946 edition of his science fiction anti-utopia entitled *Brave New World,* Huxley, notes: "It is only by means of the sciences of life that the quality of life can be radically changed. The sciences of matter can be applied in such a way that they will destroy life or make the living of it impossibly complex and uncomfortable; but, unless used as instruments by the biologists and psychologists, they can do nothing to modify the natural forms and expressions of life itself."[1] Some of the long-term trends in research in genetic engineering lead one to recall these words.

The origins of genetic engineering are usually traced to 1972 when a group of scientists at Stanford University (the United States) led by Paul Berg, subsequently a Nobel Prize winner, obtained the first hybrid DNA molecules. The methodology developed for this purpose was based on the use of enzymes which first cut DNA molecules in rigorously defined places (restrictase) and then spliced individual sections of the DNA together into a single whole (ligase). Generally genetic engineering is a system of experimental methods for creating artificial gene structures described as recombinant (hybrid) DNA molecules.

[1] Aldous Huxley, *Brave New World,* Harper & Brothers Publishers, New York and London, 1946, p. XII.

Academician Alexander Bayev observed that the emergence of genetic engineering "has not brought about a new vision of biological phenomena, new cognitive ideas or a need to radically revise existing conceptions. . . But while our earlier understanding of the nature of heredity and of its problem area has remained unchanged, possibilities for penetrating into the depth of phenomena have altered radically; a key was found to a firmly locked door, as it were, and for a time research developed freely under its own momentum. Surprisingly as it is, such a relatively modest methodological advance as the discovery and use of the restrictase and ligase enzymes has greatly expanded the scope of experimentation."[1] Let us now recall what has already been noted concerning the dialectical correlation of ends and means in human activities. The appearance of a new means for experimental and practical influences on those structures that regulate the growth and development of organisms makes it possible to envisage an entire spectrum of new objectives in a more or less distant future. Many of them, moreover, call for discussions not only with regard to their feasibility but in terms of their socio-ethical justification as well. But even that is not the end of the matter, for the development of genetic engineering has also raised the issue of ethical regulation of experimental activities generally.

The prospects being opened by genetic engineering are proving to be quite ambivalent. It may soon help the human race to obtain unlimited quantities of currently scarce medicines (insulin, the human growth hormone, interferon, numerous antibiotics, etc.). Its methods will help achieve the cherished goal of all plant growers, namely, to endow agricultural plants with such properties as resistance to disease, parasites, frost and drought, and an ability to absorb nitrogen from the air thus dispensing with the need for petroleum and natural gas to produce expensive nitrogeneous fertilizers. Finally, genetic engineering can do away with hereditary diseases by replacing pathological genes with normal ones.

Academician Bayev notes that over an extremely short period

---

[1] A. A. Bayev, "Sovremennaya biologiya kak sotsialnoye yavleniye" (Modern Biology as a Social Phenomenon), in: *Voprosy filosofii,* No. 3, 1981, p. 19.

of time genetic engineering has reached the stage of industrial applications and of practical applications more generally. While the practical applications of findings of basic research usually requires many years and often decades, Bayev estimates that in the case of genetic engineering it takes under ten years. A whole DNA industry has already been emerging since 1976 in the United States, France, Great Britain, Switzerland and other countries as firms were established to develop production processes based on genetic engineering techniques.[1]

We thus see that the potential of genetic engineering both in basic sciences and in many other fields, including practical applications, is truly unlimited. Much is being written about this today, although the forecasts being made are not always reasonably circumspect as the emphasis placed on the possible dangers is inadequate.

And yet it is extremely important to consider most seriously the potential dangers posed by genetic engineering to man and the human race. The manipulations on which its methods are based touch on the most intricate mechanisms of genetic processes of self-regulation and ultimately of life itself. Accordingly, molecular biology has reached the edge of a frightening precipice. Even simple carelessness on the part of experimenters or incompetence with regard to safety measures may produce irreparable consequences and therefore constitute a serious threat to the entire human race. Still greater damage can ensue if its methods are used by ill-intentioned persons or for military purposes. The global nature of the current danger derives above all from the fact that the organisms most frequently used in experiments are extremely widespread and are able to exchange genetic information with animals of the same species living in wild conditions. Such manipulations may produce organisms with genetic characteristics that have never existed before and that do not derive from evolution.

This has produced a very serious concern on the part of scientists that has resulted in heated debates concerning the permissibility of experiments in genetic engineering and the conditions under which they may be performed. The discussion was initiated by the concern of a number of researchers over planned ex-

[1] *Ibid.*, pp. 21-23.

periments in introducing the cancer virus SV-40, which produces turmours in mice and hamsters, into *escherchia coli* bacteria that permanently inhabit man's intestines. Both the virus and the bacteria are normally harmless to man, but concern developed that in the course of time the bacteria could spread outside experimental facilities into live human cells. It is this that led Berg, in whose laboratory these experiments were scheduled to be performed, to temporarily postpone them.

Attempts to estimate the probability that a "cancer epidemic" could develop were unsuccessful because the needed information was unavailable. This is why a special scientific conference on nucleic acids was convened in 1973 at the initiative of Berg and a number of other prominent scientists. Its basic outcome, in the words of Berg, was a recognition of how little scientists knew. Conference participants decided to address a request to the National Academy of Sciences of the USA that a comprehensive study be carried out of the possible dangers of experiments with recombinant DNA molecules and that their appeal be published.

That letter was published in the popular journal *Science*. This led to unanticipated consequences as the authors were inundated with requests for help in carrying out precisely those experiments that the conference had viewed as potentially most dangerous.

This was the situation that persuaded a group of pioneering researchers in genetic engineering led by Berg in July 1974 to address an appeal to scientists throughout the world for a moratorium on scientific research in the two most dangerous areas pending the convening of an international conference.[1] The first area concerned experiments in introducing genes of oncogenic animal viruses and toxins into bacteria. The second concerned the cloning of genes of higher organisms in bacteria. That urgent appeal for regulating scientific activity met with the support of many scientists throughout the world. In Great Britain a commission was established, headed by Professor William Ross Ashby, for examining the dangers of experiments in genetic engineering. Its work resulted in a prohibition of such experiments by Great Britain's Medical Research Council.

---

[1] See: *Science*, Vol. 181, No. 4105, 21 September 1973, p. 1114.

Such an appeal to researchers for suspending extremely promising work in a rapidly developing field of knowledge was unprecedented. True, in the late 1930s a group of nuclear physicists led by Leo Szilard addressed an appeal to colleagues throughout the world to refrain from publishing findings that could be used by nazi Germany for increasing its military power. At that time, however, it was not possible to put that appeal into practice fully and in due time. This time the constraints were to apply not to publication but to the research activities themselves.

It is interesting to note that a year later Berg observed that when publishing their appeal for a moratorium its authors were not thinking of the social responsibility of scientists: "We just felt this was a way of telling other people the way we felt about it and asking them to think about it and hold off."[1]

In commenting on the appeal Academician Engelgardt noted: "It is difficult to say whether all the motives that produced such an address are reflected in its text. But there is no question that ... the attention that was thus drawn to the problems mentioned in the scientists' appeal is entirely timely and necessary... One thing is clear already now, that the principal threat lies not in the experiments as such but in the possibility that they may be manipulated by unconcerned and irresponsible persons or else by ill-intentioned individuals. It is, above all, against these dangers that the thrust of our efforts should be aimed."[2]

The appeal for a moratorium produced an unprecedented response among wide sectors of public opinion. As Bennett and Gurin observed, it was for the first time in the history of science that a purely scientific problem relating not to scientific applications but to the appropriateness and legitimacy of the very continuation of research in this area became the subject of wide discussion.[3] Many articles appeared in American and British periodicals to the effect that if scientists themselves decided to

---

[1] William Bennett, Joel Gurin, "Science That Frightens Scientists: The great debate over DNA", in: *The Atlantic,* Vol. 239, No. 2, February 1977, p. 49.

[2] V. A. Engelgardt, "Dve storony molekularnoi inzhenerii" (Two Aspects of Molecular Engineering), in: *Za rubezhom,* No. 36, 1974, p. 20.

[3] See: William Bennett, Joel Gurin, *op. cit.,* pp. 43-44.

renounce certain types of experiments then all research in genetic engineering had to be dangerous.

The moratorium was honoured almost universally for eight months, until the end of February 1975, when an international conference was convened in Asilomar, California, at which 140 scientists from 17 countries including the Soviet Union reviewed the scientific advances in research on recombinant DNA molecules, discussed a number of social and ethical aspects of experimentation in genetic engineering, ways of avoiding potential biological hazards and conditions for lifting the voluntary moratorium on the two most dangerous categories of experiments.

The atmosphere in which the conference took place reflected the fact that the scientists had to make a decision that would influence science policy, the distribution of funds and largely predetermine the further development of genetic engineering. In view of the heightened interest displayed by public opinion and government circles, recognizing particular types of experiments as dangerous would have automatically changed research conditions for the worse and would have even entailed complete termination of certain studies which had unquestionable scientific value. In describing the behaviour of the participants in the conference, Berg observed that each of them sought to draw the line around his own research subject, allegedly "pure" and "harmless", and to show that it was the research of others that was dangerous and had to be prohibited.

In reporting on the discussions at the Asilomar conference and commenting on its outcome the journal *Science* noted: "Like the moratorium that preceded it, the conference's statement has the power of moral censure only, but the guidelines it proposes will probably be followed closely by the national bodies in each country responsible for framing the relevant regulations.

"Just as the moratorium seems to be unprecedented in the history of science, the action of the conference is a rare, if not unique, example of safety precautions being imposed on a technical development before, instead of after, the first occurrence of the hazard being guarded against.

"The conference's decisions were reached in the explicit awareness that... if the scientists present failed to regulate themselves in an evidently disinterested manner, others would do so for them."[1]

According to *Science,* the conference's organizing committee's principal opponents were James Watson and Joshua Lederberg. Acting as a pair of *enfants terribles,* these Nobel Prize winners, even though lacking an agreed position, continually sought out problems in its recommendations and sowed discord among young scientists. Lederberg described as "too definite" a document prepared by the conference working group where possible experiments were categorized into six classes corresponding to the anticipated degree of danger and procedures were recommended for each that, according to the group, reduced the biological hazards to a commonly acceptable level of risk. Watson openly stated that the moratorium had to be abandoned. In signing the letter concerning the moratorium, he said, he had thought some time would elapse and he would perhaps learn of something that would frighten him. But as head of laboratory on oncogenic viruses, he added, he concluded that he was working with something that was more dangerous than anything he had heard at the conference.

It was precisely on the issue of categorizing experiments in terms of their relative danger that the main problem emerged, as it became evident that there was no agreement concerning the manner in which danger could be defined. Lederberg stated that if the proposed classification was to become the basis of legislation a greater degree of certainty was needed as to its correctness. Watson insisted that scientists could not measure risk and that there were accordingly no grounds for discontinuing research activities for the sake of something that could not be measured.

Such uncertainties, of course, made it difficult to arrive at clear decisions. Nevertheless, the conference did achieve some positive results. It is true that as one analyzes its decisions it is evident that the Asilomar conference was only concerned with avoiding dangers produced by possible errors and carelessness on the part of experimenters and did not address itself at all to the major issue facing each researcher—namely, who and in serving what objectives will make use of the results of their work. But it cannot be denied that on the whole the conference

[1] *Science,* Vol. 187, 14 March 1975, p. 931.

had played a positive role. Even though some of the scientists did not accept the need for any form of self-regulation, its main achievement, in Berg's opinion, was that it "raised the level of discussion in the field". He added that no one would go into that field, without thinking about the risks and benefits and this could not have been said eight months earlier.[1]

Characteristically in a letter to the Editor of *The New York Times,* Lederberg thought it necessary to refute a news item on his differences with the Asilomar conference stating that he viewed the precautions as being virtually inoperative because of the difficulties associated with determining precisely the risk in specific experiments. "My own view," he wrote, "is almost the opposite of what was asserted. My chief concern at the conference was that a set of precautions that are entirely appropriate for certain risks might be prematurely rigidified into a set of bureaucratic regulations that might be very readily enforced beyond the domain of their reasonable application. I am wholeheartedly in support of the spirit and intentions of the conference report but was unwilling to put my name to a document that left many important questions for future determination and whose tone seemed to invite the bureaucratic rigidity just mentioned."[2]

The conference's Summary Statement noted that impressive scientific advances had already been made in research on recombinant DNA molecules, and that new prospects were thus opening for a deeper understanding of fundamental biochemical processes.[3] Methods for working with recombinant DNA molecules promised to evolutionize the practical aspects of molecular biochemistry. While the new methods had not yet been applied in practice, there was every reason to believe that their enormous practical significance would be eventually revealed.

Many scientists were particularly interested in the question of whether there should be a lifting of the suspension of research in certain areas in accordance with the appeal contained in the July, 1974 letter of the Committee on Recombinant DNA Mo-

---

[1] *Science News,* Vol. 107, No. 10, March 8, 1975, p. 156.

[2] *The New York Times,* March 11, 1975.

[3] See: "Summary Statement of the Asilomar Conference on Recombinant DNA Molecules", in: *Proceedings of the National Academy of Sciences of the USA,* Vol. 72, No. 6, June 1975, pp. 1981-1984.

lecules of the National Academy of Sciences of the USA; in the event of its discontinuation, they wanted to know the procedures of scientific research that would minimize the risk for laboratory staff, the population at large, and animals and plants.

In its Summary Statement the conference touched upon the problem of new methods that would make it possible to combine the genetic information of quite distinct organisms and thus rose problems involving many unknown factors. Even with the existing level of research the evaluation of potential biological hazards was extremely difficult. Thus the conclusion was that considerable caution was needed in such research activities. Nevertheless, the conference's participants agreed that a large part of the work in constructing recombinant DNA molecules could be carried out provided that sufficient safety measures—primarily biological and physical barriers to safely isolate newly created organisms—were employed. Moreover, precautionary standards were to be higher at the initial stages of research, and were to be revised as the methodology developed and as estimates of anticipated risk changed. It was also recognized that in some experiments the potential risk is so great that they were not to be performed with existing safety measures. In a more distant future, applications of experimental findings in industry, medicine and agriculture could cause problems. At the same time, future research and accumulated experience may show that many of the potential threats are less serious or probable than they appear to be today.

The conference formulated the following principles basic to its recommendations and conclusions: (1) in designing experiments it is essential to consider questions of safety; (2) the effectiveness of safety measures must correspond as much as possible to estimates of anticipated risk. Accordingly, when agreement has been reached concerning the risk scale, a corresponding scale of safety measures will have to be developed. Periodically the two scales will have to be revised. One would hope, the Summary Statement added, that a combined use of both formal and informal national and international information channels will make it possible to develop a reliable method for matching potential biological hazards with corresponding safety levels.

Permissible experiments were divided into three categories in terms of associated risk, ranging from experiments with minimal risks to highly dangerous experiments. In addition, a fourth group of experiments was defined that are at present so dangerous that they should not be carried out at all pending the elaboration of corresponding safety measures. In the opinion of the participants, the principal way to avoid known and possible dangers is to construct genetically new bacteria and viruses possessing a limited ability to multiply outside laboratory conditions. The conference's Summary Statement also placed considerable emphasis on the need to fully inform all personnel associated with experiments concerning the level of risk involved and to organize personnel training in the safety measures required for these experiments.

The call for a temporary moratorium and the Asilomar conference were culminating points in the development of events associated with genetic engineering. After the conference, however, discussions in scientific circles flared up with a new force, for many researchers viewed the conference's decisions as excessively strict and lacking a rational basis.

It should be noted that the conference's ban on certain experiments did produce some reorientation in genetic engineering research. In particular, emphasis was placed on the creation of weakened vectors, that is of transmuted microorganisms lending themselves to laboratory experiments but unable to survive outside laboratory conditions (they do not settle in human intestinal tracts or semen and are easily destroyed by ordinary detergents). This makes it possible to overcome one of the basic obstacles to research in recombining DNA.

In 1975 a Committee of National Institutes of Health (the United States) sought to rely on the Asilomar conference's fundamental guidelines in developing acceptable instructions that would make it possible to control the use of methods for manipulating the genes of living organisms, and, in particular, the use of weakened microorganisms in many types of experiments. The Committee established a number of levels of normal physical protection, ranging from standard microbioloical methods to the use of specially designed equipment using reduced air pressure, insulated through air buffers and with showers at each entry point.

On the whole, however, that Committee's standards were substantially less rigorous than the recommendations of the Asilomar conference, and they were sharply criticized by a number of scientists. In particular, J. King, a professor at the Massachusetts Institute of Technology specializing in molecular biology, noted with irony that judging by the standards proposed by the National Institutes of Health their aim was to protect geneticists rather than the public. He stressed that the results of experiments in genetic engineering could be utilized by the United States government for military purposes. He also noted the possibility that some human or animal gene, introduced into bacteria, could be capable of producing a certain type of protein which, upon absorption by a human organism, may activate the system of immune protection, leading to the production of antibodies. The latter, in turn, may become the source of illnesses that were previously unknown.

And while some participants insisted that the probability of such phenomena was very low, advocates of rigorous regulation of research activities maintained that, since what was involved was a threat to the human organism's genetic constitution, that probability cannot be neglected. As a result the National Institutes of Health had to make their guidelines more rigorous. The new list of requirements published in late 1975 was even more rigorous than the Asilomar conference's recommendations.

This was followed by re-equipment of laboratories in conformity with the new standards, as universities began to plan new research activities in genetic engineering. In Cambridge, Massachusetts, however, where both Harvard University and the Massachusetts Institute of Technology are located, events unexpectedly took a new turn as the debate was joined by representatives of municipal authorities and public opinion concerned over press reports on the possible dangerous consequences of research in genetic engineering.

The views of Cambridge scientists were divided—some were convinced such experiments were completely safe, while others demanded banning them. Decision-making was in the hands of proponents of the ban. A special committee was formed that included public figures and representatives of city authorities; not a single scientist had been invited. What scientists had sought to avoid had occurred: the safety of biological experiments was evaluated by persons who were poorly qualified. And while the committee's conclusions were not mandatory in nature and could only influence the decisions of Cambridge's municipal authorities, it was clear to all that its findings would influence public opinion in many countries.

Having twice postponed granting the permission to carry out dangerous experiments, the municipal authorities finally authorized them in February 1977 with the condition that genetically weakened microorganisms would be used and all the guidelines developed by the National Institutes of Health would be adhered to. Experiments of the most dangerous fourth type were not permitted.

Subsequent discussion reflected increasingly the view that national legislative measures were urgently needed to regulate research in recombinant DNA. Senator Edward Kennedy and Congressman Paul Rogers proposed bills that the US Congress began to discuss early in 1977. Both bills envisaged extremely strict forms of government control over research activities.

Scientists as well as congressmen participated widely in these debates. Many biologists who feared excessively rigorous forms of regulation came to oppose the bills. They stressed that because recombination processes frequently occur in Nature they do not constitute something fundamentally new. Occasionally, however, it was evident that some of the scientists opposing the bills were defending not only the principle of freedom of research but also their own not altogether altruistic objectives.

The American Society for Microbiology voiced its opposition to the draft legislation and was supported by 20 more scientific societies in the United States and by many leading biologists, some of whom enlisted the support of senators and began to organize a powerful Congressional lobby. In view of this, Senator Edward Kennedy announced in September 1978 that he was withdrawing his bill, proposing instead the creation of a fact-finding commission to examine the nature and the potential hazards of experiments with recombinant forms of DNA.

Participants in these discussions recognized that what was involved was not only the future of genetic engineering itself but also the general role of science in society and the principles of social control over science's development. "The rancor generated by the dispute hardly makes sense as a specific response to this particular line of experimentation. In calling scientists to account for their work with genetically modified bacteria, the public and its representatives seem to be seizing an opportunity to establish a more general principle. A recondite area of modern biology has become a metaphor: the ultimate question is not whether bacteria can be contained in special laboratories but whether scientists can be contained in an ordinary society."[1]

The intensity of these discussions has by now noticeably dwindled; still, it would be incorrect to say that the situation in genetic engineering has become fully clear and unambiguous. As before, there are problems requiring immediate solution, and many compare research in recombinant DNA with the ensuing dangers to a genie released from the bottle. Numerous sensationalistic articles and reports by journalists and radio and television commentators to the effect that a new Frankenstein-like monstre may soon crawl out of laboratories have stirred world public opinion and alarmed a number of scientists in the West. More recently, however, they have become concerned with still another problem, namely, a possible decline in funding because of the pressure of public opinion and of legislative decisions of the administrative authorities. This, many scientists believe, could impede the development of science in general. Accordingly, the growing movement for halting dangerous experiments is accompanied by an intensive search for acceptable forms of continuing work in genetic engineering. In such a context an important role is played by legal and legislative aspects of the problem. Proposals to make forms of control over research in some cases less rigorous and to revise individual rules in the light of the experience accumulated in recent years are now often heard. These proposals have been criticized by some scientists; others, however, consider that current rules are unjustifiably harsh.

James Watson, a Nobel Prize winner, has even called for an end to government rules regulating recombinant DNA research. He described these rules as unneeded "nonsense" and said "there is no evidence" that recombining genes can do anyone any harm. The enormous amount of time and dollars spent on maintaining restrictions is, in his opinion, unjustifiable. The restrictions have become a "disaster", Dr. Watson said, and the public has been

[1] William Bennett, Joel Gurin, "Science That Frightens Scientists...", *op. cit.*, p. 62.

misled to fear "madman scenarios".[1] It may be recalled that at some point Dr. Watson has been equally vigorous in calling for restrictions on experiments in genetic engineering.

Today some doubts have even been expressed as to whether the movement for the moratorium and the decisions of the Asilomar conference were motivated entirely by noble intentions—in view of their subsequent repercussions both among scientists and more generally. It is difficult to answer this question unequivocally, for in the present world, in which a sharp political and ideological struggle is taking place, there is probably no problem or action that, conditions permitting, could not be misused by reactionary social forces in some way or another. Accordingly, even the noblest and most humane proposals are often distorted and used for directly opposite purposes.

In noting the negative aspects of the phenomenon that we are considering we criticize as negative the actions of Western governments and propaganda agencies, but not the movement of scientists itself. One cannot doubt the sincerity of the original motives of scientists who initiated the movement for a moratorium on certain types of experiments with recombinant DNA. It is also true, however, that the subsequent behaviour of these scientists has not always been unambiguous and consistent.

It is also important that discussions concerning genetic engineering have compelled many scientists to take a new look at their relations with society and their social responsibility. In particular, David Baltimore, an American biochemist and Nobel Prize winner, has stated that these discussions led him to realize that "all of the magnificence of modern science has not brought qualitative changes in people's feelings about the world, has not eradicated poverty". With regard to the discussions themselves, he added: "I think that we had a responsibility to do it and I don't see how we could have avoided that responsibility. But we haven't provided what we wanted to provide, which was a smooth transition from a kind of laissez-faire policy in science to a rational evaluation of hazards."[2]

This is also confirmed by the position of Gerard Weissman, an

---

[1] *International Herald Tribune,* December 20, 1977.

[2] Quoted in: William Bennett, Joel Gurin, "Science That Frightens Scientists. . .", *op. cit.*, p. 62.

American biologist, in an article where, in his own words, he defended a point of view that is both liberal and risky: one cannot place limits on the curiosity of scientists. "Truth at any cost", that is, the unconstrained study of anything and at any time is a major cultural value of civilization. In opposing the restrictions on research in recombining DNA, he writes that prohibitions on lines of research that may result in the creation of chimeras are based on the assumption that scientific research can produce only undesirable consequences. Of course, Weissman greatly oversimplifies the position of his opponents. No serious person would argue that experiments can produce only negative consequences. The problem is to anticipate—and, whenever possible, preclude—such consequences. This problem is also relevant in the context of socio-ethical control over research in genetic engineering.

Today, when intensive work in genetic engineering is being carried out in many laboratories throughout the world, the debates concerning research in recombinant DNA may be substantially viewed as a thing of the past. This may be attributed to a number of causes: the fact that a system of safety measures in carrying out experiments has been developed; that it has become possible to assess the risk factor somewhat more accurately; and, finally, to the fact that, having focused their attention on their own problems, scientists discovered that both discussions and especially decision-making on key issues were evading their own control. This has produced some new fears, as to some extent a concern over incompetent forms of intervention in research activities has replaced their original concern with the potential dangers of research itself.

Thus, while many problems still remain, the moratorium on experiments may now be viewed and analyzed as, in a sense, a fact of science's history. In such a context, it is important to note that it is a fact of not only what is known as the "external history of science" which examines, among other things, social relations in science, i. e., the history of scientific views, conceptions and methods of research, etc.

Nevertheless, there are still different interpretations of this fact, as has been revealed, for example, at the Third All-Union Conference on Philosophical Problems in the Natural Sciences held in Moscow in April 1981. The ethical problems of science, including those associated with the moratorium, were considered in many presentations. And that is an indication of the current relevance and serious nature of these problems and evidence of the widening philosophical scope of issues raised by the natural sciences.

In his presentation Academician Bayev noted that the call for a voluntary temporary renunciation of research had marked the beginning, mostly in the United States, of a campaign against genetic engineering, instantly joined by the sensationalistic American media. On the whole, however, "the social impact of genetic engineering involved ethical, philosophical and social problems, and even though it was based on unfounded premises, its very existence is evidence of the important role that biology is beginning to play in modern society."[1]

Academician Engelgardt presented a different evaluation of these events. He noted that while the moratorium was opposed by advocates of "freedom of scientific research", common sense prevailed, and as of this stage corresponding rules of work have been adopted in most countries, sometimes even in the form of legislation. Thus the joint efforts of scientists have prevented a major threat that could have reached crisis proportions. The Asilomar moratorium "can justifiably be viewed as a prototype form of the expression by scientists of their responsibility in confronting a danger that could have become a national calamity and a crisis"[2].

The above views obviously differ, and the constructive distinction between them, in our view, could provide a convenient point of departure for further analysis.

In considering both views one may note the following. It is quite possible, of course, that the call for the moratorium could have been prompted not so much by the real dangers associated with such experiments as by such extraneous circumstances as self-promotion or a wish for greater prestige in the eyes of public opinion. But even if this were so, the fact that the appeal was addressed to public opinion is revealing, for efforts were made to win the latter's attention not by promising certain benefits in

---

[1] *Dialektika v naukakh o prirode i cheloveke* (Dialectics in Sciences of Nature and Man). Ed. by I. Frolov, Nauka, Moscow, 1983, p. 116.

[2] *Ibid.,* p. 128.

this particular sphere of research but by warning of possible dangers. This means that expressions of the feeling of social responsibility are not only a socially acceptable but a socially recognized, and possibly even a socially promoted form of behaviour on the part of scientists.

Today there are also sufficient grounds for asserting that, on the whole, the potential dangers of these experiments were exaggerated. But that was hardly evident at the time. Current scientific knowledge concerning the safety of some experiments and the dangers of others is itself a result of research activities. For both the categorization of experiments based on their relative hazard and the development of methods for producing weakened viruses that can exist only in artificial laboratory environments represent new scientific data and new methods of experimenting arrived at as a result of the moratorium and of subsequent discussions that stimulated research in this field.

I would like to place particular emphasis on this point, which I view as fundamental, for it shows that in actual fact the scientists' feeling of social responsibility, often described disparagingly as both ephemeral and negligible, may operate as an active and effective factor in scientific cognition. It follows that this social responsibility is a phenomenon associated not only with the participation of scientists in public movements but that it may also find immediate forms of expression in the course of a scientist's own cognitive activities. The above applies, of course, not only to the moratorium or genetic engineering but to a wider context as well.

If we discard the constraints of a conception of scientific knowledge as a blind adherence to a rigid and entirely mechanical "internal logic" of the development of science, and if we fully recognize that cognition involves solving problems of deep and immediate interest to scientists precisely in the context of the totality of social problems, then it will not be difficult to agree that a problem's social significance often serves as an important stimulus that reinforces the interest displayed by scientists. In the case of research seeking to identify the dangers of recombinant forms of DNA, it is precisely their social significance that determined the scientific interest towards the problem.

It follows that the social responsibility of scientists and of science is not something extraneous that is unnaturally imposed on scientific activities as such. On the contrary, it is an inherent component of scientific activity that may exert a palpable influence on the problem areas and direction of scientific research.

In an overall evaluation of the discussions which have taken place on the subject of genetic engineering, one should note that they have provoked a marked increase in the interest displayed by wide segments of public towards the socio-ethical aspects of modern science's development and towards questions relating to the social responsibility of scientists. But since an acceptable practical solution to the problems raised by the moratorium was achieved, that interest naturally began to subside.

There is no guarantee, however, that new ethical challenges will not emerge in some field of knowledge in the near or distant future. The extent to which science will be prepared to resolve such issues will largely depend on further advances in serious analytical studies in the ethics of science. In this context, the significance of discussions on the socio-ethical problems of genetic engineering extends far beyond one specific episode in the history of a specialized branch of biology. The experience that has been gained by the scientists who for the first time found themselves in a complex situation that was not amenable to any simple approach or solution will continue to play a positive role in the future as well.

Many Western scientists face a sharp and complex moral-ethical dilemma. Their responsibility to society has prompted one type of behaviour, namely, informing public opinion of the possible hazards of planned research and developing special precautionary measures for carrying out experiments. Their responsibility towards science, on the other hand, called for completely different actions. The point in question is not only that scientists have a responsibility before science for the novelty and accuracy of the knowledge that they convey, but, above all, that a real danger had emerged of creating a precedent as regards incompetent and bureaucratic forms of intervention in research. Subsequently this could have spread beyond the sphere of genetic engineering and entailed unpredictable consequences. The scientific community's behaviour in that tense situation was by no means unambivalent, and it has become quite clear that the existing mechanisms of self-regulation in science, based on ethical criteria that have developed spontaneously, are not always satisfactory, if effective at all. For in bourgeois countries the interests of science and those of society are most often in a state of dramatic conflict which renders useless the mechanism of self-regulation.

With regard to the lessons of the Asilomar moratorium one should also note a point that remained unnoticed in the course of the discussions: in one important respect, the position of the advocates of "absolute freedom of research" was inherently contradictory. By denying *a priori* the potential danger of experiments, and, in particular, the possibility that harmful hybrid forms could appear, they by the same token indirectly acknowledged the impracticability of the experiments—for one cannot hope for significant results in research if one does not expect it to produce viable hybrid organisms. Yet it is precisely the risk of producing viable hybrids with unfamiliar characteristics that became the basis of all subsequent assessments of the potential hazards of the experiments.

In socialist countries genetic engineering developed in a moral atmosphere that was quite different. A leading Soviet specialist in genetic engineering, Academician Bayev, has rightly noted that "people in the Soviet Union have neither fears for the future nor apprehensions that some powerful and blind forces might direct research in genetic engineering onto an evil course in spite of the intentions and wishes of the people. We are persuaded that in this respect reason and goodwill will prevail—at least in this socialist country."[1] At the same time, both in the USSR and in other socialist countries discussions of the socio-ethical and humanistic problems of genetic engineering do take place, often at the initiative of the specialists themselves (Bayev, Yuri Ovchinnikov, Engelgardt) and others.

Research specialists throughout the world can hardly limit themselves to discussions of the cognitive and technological problems of genetic engineering, in the belief that everything else, including the possibility of the development of biological weapons, depends not on them but on political leaders, social relations, and international agreements. Such a position is dan-

---

[1] A. A. Bayev, "Sotsialnye aspekty geneticheskoi inzhenerii" (Social Aspects of Genetic Engineering), in: *Filosofskaya borba idei v sovremennom yestestvoznanii* (Philosophical and Ideological Struggle in Modern Natural Science), Moscow, 1977, p. 146.

gerous, for it isolates scientists from mass public movements. It may disorient those scientists who oppose inhumane uses of the achievements of all modern sciences and whose protests play a major role in the struggle for peace and disarmament and against imperialism and militarism.

Thus, in modern science socio-ethical problems arise both with regard to each scientific discovery and research problem as well as with regard to science's overall objectives. Current discussions on problems associated with regulating research in genetic engineering should not be viewed as something temporary, transitory or accidental in the development of science. They are becoming an inherent part of scientific activity itself, and this points to the growing role of science in society and in the lives of all individuals.

The most acute socio-ethical problems that have emerged in connection with the prospective development in genetic engineering can, of course, be resolved on a genuinely humanistic basis that attaches priority to the benefit of man. At the same time, their solution should not preclude new approaches in the cognition of Nature and man. In the final analysis, it is cognition that serves man's wellbeing and offers one of his major hopes for the future. The human race must rise to a new level of social and moral development in order that this hope may be realized.

This is why more attention should be paid to science's socio-ethical problems which link the human race to science and scientists in a thousand ways and make the latter responsible for man's future. There is a need for comprehensive scientific discussions, comparison and opposition of ideas, and for a creative dialogue between scientists adhering to different philosophical and socio-ethical views. It is relatively difficult to define the risk factor involved in individual research activities and correlate it with the possible social benefits in each particular case. But it is precisely for this reason that science's socio-ethical problems must always be emphasized not only by experimental scientists, but also by philosophers specializing in ethics, jurists and other scholars. These problems should also become a matter of permanent concern to political leaders and to world public opinion.

It has already been noted that a number of Western scientists who had enthusiastically addressed the ethical aspects of genetic engineering are now stating that discussing them widely is unnecessary and may even be harmful. They claim that the associated dangers were exaggerated by the mass media and that any prohibitions or constraints in this area might hold back progress in scientific research in this new field of cognition that is of great importance for mankind.

Yet such a negative position is misconceived. Genetic engineering has demonstrated to scientists and public opinion more graphically than any previous development (including discussions of the dangers of research in nuclear physics) the need for controls (both social and ethical) over all aspects of scientific activities that may pose a direct threat to man. So far matters have generally been limited to a "drama of ideas" and, to an even greater extent, to sharp verbal exchanges. But all major developments still lie ahead—both the good ones (for genetic engineering holds out the promise of future miracles both in agriculture and in medicine) and the bad ones (although it is true that many specialists are consoling us that new weapons of mass destruction that may be created with the help of gennetic engineering will not be more destructive than nuclear missile systems). This remains, of course, to be seen.

Even though some ground rules in genetic engineering have now been adopted, it is hardly appropriate to minimize the potential dangers of uncontrolled research in and technological applications of genetic engineering in capitalist society, especially with regard to the development of biological weapons, various neo-eugenic projects and cloning.[1] We take a negative view of all attempts to somehow bypass already adopted agreements and prohibitions under the pretext of concern for scientific progress and to describe discussions of science's ethical problems and of the socio-ethical regulation of work in genetic engineering as "harmful" and as impeding progress in scientific research. Sci-

[1] In recent years American experts have stated that the methods of genetic engineering cannot produce a biological weapon that would be superior to weapons developed by other methods. Also under discussion is the idea of using genetic engineering to develop vaccines against bacteriological weapons. See: E. Gaissler, "O zloupotreblenii resultatami issledovanii v oblasti molekulyarnoi i kletochnoi biologii: fakty i mify" (On the Misuse of Research Findings in Molecular and Cellular Biology: Facts vs. Myths), in: *Mir nauki,* No. 1, 1982.

entists in socialist countries support all proposals and activities of scientists in capitalist states aimed at placing numerous studies, and in particular those in genetic engineering, under effective social control. We have therefore taken a positive view of their proposals for a moratorium on certain types of research. While we recognize the shortcomings and contradictions inherent in such measures, and, in particular, the utopian nature of the idea of self-regulation of science based exclusively on moral and ethical norms, we also consider that such forms of protest against possible inhumane uses of science should not be discounted.

It is important to show greater wisdom in discussing the possible outcomes of such research and its practical applications and to emphasize the importance of their *social contexts*. It should be stressed that the enthusiasm of those persons who seek to solve social and human problems through genetic control can hardly be shared. It is only at first glance that this approach may superficially seem to be the quickest and most reliable method. In actual fact, however, it raises new problems, even more formidable than those that the proponents of genetic control seek to resolve. Whenever scientists' humanism loses a concrete-historical and socio-ethical perspective it becomes either something ephemeral or even—in view of its abstract nature—potentially its opposite, namely, anti-humanism.

Discussions concerning the ethics of genetic control often turn to relatively remote and sometimes simply utopian prospects that might emerge with the development of genetics. The urgency and intensity of these discussions is attributable not so much to the feasibility of these remote goals as to the fact that they force us to approach in a new way or to perceive more intensely such eternal problems as those of man's freedom, his potential, and his mission. In particular, the prospects being opened by genetics force us to reflect on whether we do wish and should wish to see human beings reproduce through cloning.

The discussions therefore are concerned not with only the future but also with very urgent contemporary matters. In this context, it is not only desirable but essential to turn to the experience accumulated in philosophical studies of these problems over the centuries. Current scientific advances again and again confront the human race with the problem of self-determination and of reconsidering its system of values, and unless man uses the experience accumulated by philosophy he may lose his orientation in responding to that challenge.

In Western literature the emphasis is usually placed on the human race's ethical, abstract-humanistic, and religious values, with religion most often being viewed as a science's moral guardian. Such an approach to the ethics of genetic control is evident in the views of Paul Ramsey, who states, with reference to the permissibility of preventive genetic (and even eugenic) control, that all scientific recommendations and any method for carrying them out must be harmonized with the principles of Christian ethics.[1] This is the sense in which he proceeds to discuss the ethics of genetic duties in arguing against Lederberg's views concerning the acceptability of cloning for man.

In elaborating a technology for replicating human beings, Lederberg seeks to show its advantages compared to genetic engineering, since the former is concerned with reproducing already known genotypes whereas the latter involves the creation of new ones, associated with dangers and risks. In both cases Lederberg accords priority to "pure experimentation" rather than to moral and ethical considerations. He takes the view that in order to carry out experiments in cloning it is sufficient to have the personal consent of those who would be genetically replicated. He fails to take into account the danger that such practices may produce an elite that would consider itself above society. Moreover, he even views this as an advantage because, in his opinion, members of such elites would be linked by close social ties, the struggle among generations would disappear, and so on.

The view that research in genetic control and corresponding practices should be considered in isolation from social, humanistic and ethical values is widely held in the West. As for Lederberg, it would seem that under the influence of the tendency to take ethical norms into account, displayed by many scientists, he has greatly altered his earlier position, as seen, in particular, in his foreword to Joseph Fletcher's *The Ethics of Genetic Control*,[2] and in his statement made during the discussion of the socio-ethical regulation of genetic engineering.

---

[1] Paul Ramsey, *Fabricated Man: The Ethics of Genetic Control,* Yale University Press, New Haven, 1971, pp. 22-59.

[2] Joseph F. Fletcher, *The Ethics of Genetic Control: Ending Reproductive Roulette,* Ancor Press, New York, 1974.

With regard to the future of genetic research, and specifically in connection with cloning, he observed in a letter written in October 1977 to Ruth Kirschstein, Director of the National Institute, General Medicine, that one of the basic objectives faced by researchers in genetics is to minimize the suffering produced by genetic factors in other illnesses. This requires a strong emphasis on developing effective methods for intrauterine diagnostics. This also includes studies in molecular and cell biology which open new prospects for the therapy of genetic diseases. Nor should one forget the prospects for developing methods of treatment that reduce suffering from such illnesses without affecting the organism's genetic basis.

Lederberg notes that there are many discussions of the allegedly existing possibility of genetic modification of human beings, for the justification of which utopian or medical considerations are advanced. He notes also the superficial nature of such discussions, for no method of genetic modification exists which would not be associated with risk for man. Moreover, Lederberg stresses that these methods are absolutely unacceptable socially, for our knowledge of man's biological potential is too limited and hence any interference in his genetic basis is morally repulsive.

Lederberg's letter was published as an appendix to *The Cloning of Man: A Brave New Hope—or Horror?* that contains a comprehensive discussion of the problems of genetic engineering.[1] The author states that discussions concerning cloning were stimulated by the publication of David Rorvik's book entitled *In His Image*.[2] Rorvik claims that a cloning of man has already been performed and that in mid-1978 that clone was already one and a half years old. In Lederberg's view, the scandalous notoriety of Rorvik's book has greatly discredited the idea of cloning. When viewed in the context of current genetic and biological research, however, cloning seems to deserve close attention and serious discussion.

When J. B. Gardon first performed an experiment in cloning a frog in the mid-1960s, even the most conservative special-

---

[1] *The Cloning of Man. . ., op. cit.*, p. 198.

[2] David M. Rorvik, *In His Image,* J. B. Lippincott, Philadelphia, 1978.

ists predicted that man, too, could be cloned within another decade. A further impulse to these discussions was imparted in 1978 by a successful operation performed by two British researchers, Patrick Steptoe, a gynecologist, and Robert Edwards, a physiologist, who artificially fertilized and subsequently implanted in uterus the egg cells of a patient at Cambridge University's clinic. That operation was followed by the birth of a girl whose photograph was published in newspapers throughout the world. She was the first human being whose life had begun in a test tube. While the method employed by Steptoe and Edwards differed from that of cloning, it also caused an extremely heated discussion to which we shall return. The need to analyze the possible consequences of such cloning is considered in the concluding chapter, entitled "Is There Nothing to Stop It?" of Ebon's book.

In considering various answers to that question, Ebon writes: "Speaking as a public-spirited citizen myself: Any effort to clone a human being, to create a man-made man, is, in my opinion, immoral, tragic and totally unnecessary."[1]

That statement is symptomatic, and while it has been made by a scientist who has not engaged in genetic research himself, it does express the general position of many geneticists. In this regard the views concerning cloning expressed by James Watson, a Nobel Prize winner, at a meeting convened by the House of Representatives of the US Congress (January 26-28, 1971) and also published as an appendix to *The Cloning of Man,* are characteristic.

Watson believes that we will soon face some difficult decisions. It is absolutely clear that the potential of abuse in cloning will outweigh the possibilities of helping childless families. It is the duty of scientists to inform as many people throughout the world as possible of the prospects and consequences of the emerging possibilities of replicating people. Today the possible negative as well as positive impications call for serious consideration not only by scientists but by the widest segments of public opinion.

New forms of international awareness and corresponding international agreements will have to be developed. Some persons

[1] *The Cloning of Man...,* p. 168.

may have doubts concerning the feasibility of such agreements or the emergence of such an awareness. But unless we tackle these problems without delay we may one day discover that we have lost our ability to decide freely, Watson concludes.

One cannot, of course, argue against the above conclusion in its general form; it is important, however, to decide on what basis and in what form it should be implemented.

In this regard, those scientists who are Marxists see a wide range of problems calling for a close analysis. These issues were addressed, in particular, at a round table held by the journal *Voprosy filosofii* where the following views were expressed. In referring to recent studies concerning the transplantation of nuclei and to the possibility of applying them to human beings (which may make it possible to artificially produce monozygotic twins) Abram Neifakh expressed the view that, barring the possibility of serious illnesses, trauma or other major irregularities, one may expect that such twins will be similar not only physiologically but also psychologically. It is true, he noted, that the human psyche depends more than does morphology on conditions of development, training and education and that therefore substantial deviations are possible. Nevertheless, if the conditions are similar one may expect that the basic psychological traits of monozygotic twins, and in particular their creative abilities, will be very similar. This, in Neifakh's opinion, suggests that there may be a new way to create the particularly gifted individuals that are needed for accelerating scientific and technological progress, developing the arts, etc. It follows that the method of transplanting nuclei, applied to human beings, would preserve exceptional genetic combinations that sometimes emerge accidentally and, naturally, vanish with the individuals' death. Such unjustifiable losses for the entire human race would then be avoided. In addition, Neifakh observed that it is not logical to refer to the method of transplanting nuclei as applied to man as eugenics, since the latter seeks to alter and improve the nature of the entire human race, while the transplantation method simply preserves what already exists and increases the number of especially gifted persons without changing their genetics. It also extends, moreover, only to selected individuals rather than the entire human race.

In this connection, however, many problems arise. Do we have the right to prolong the lives of genetic combinations that would vanish naturally with the death of its carrier? Do women have the right to bear children genetically not their own? How should such "twins" be brought up and educated? In particular, would it be proper to prepare them from the outset for specific occupations? Finally, can it be ruled out that the method of transplanting nuclei could be employed to the detriment of the human race? Would we not create some elite which would displace persons born "in the usual manner"? While Neifakh considers that these concerns are justified, he believes that it is still difficult to decide what is worse—to hold back progress or to burden the world with a new source of major anxieties. A certain danger does remain. At the same time, the positive aspects of the method of transplanting nuclei are also apparent, even though its application is still a matter for the future.

A largely opposite point of view was repeatedly expressed by Nikolai Dubinin. To him, the selection techniques applied to animals hold no promise whatsoever for the genetic improvement of man. It is only natural that individuals differ genetically. But should one seek to destroy that diversity? It is entirely possible that this diversity constitutes the very basis of genetic stability and of the human race's potentialities generally. On the other hand, if we were to invent some ideal human personality and recklessly undertake to remake all persons to conform to that ideal this would result in a dismal failure, for we are absolutely ignorant of the genetic foundations of the human personality. Far from being an attempt to avoid the problem of heredity, a clear understanding of the impermissibility of uninformed interference into human heredity is, in Dubinin's view, an expression of a scientist's social responsibility.

Dubinin stated that views such as those defended by Neifakh are based on an exaggeration of the role of genetic programmes to the detriment of social hereditary programmes. In his opinion, Neifakh's attempt to consider the consequences of an abstract possibility of producing human beings from somatic cells has substantial shortcomings, for it is based on an oversimplified interpretation of the human race's future and of the role of exceptionally brilliant persons in history. Attempts to create groups of individuals who would possess inborn biologically determined capacities for highly effective forms of mental activity threaten to seriously disrupt the entire fabric of society. Whatever the original purpose of such specialized groups, their emergence would eventually result in both the creation of a distinct caste and the standardization of human beings. Scientific development is, of course, unavoidable, and future generations will achieve unimaginable power over Nature, but this will take place without the creation of biologically differentiated groups.

Subsequent discussions produced a further development of the approaches presented by Neifakh and Dubinin and also brought to light additional considerations relating to the prospects of human genetics. Leontiev observed that, unfortunately, in many studies in human genetics careful analysis of psychological factors was replaced by their superficial interpretations. For example, there are frequent references to psychological data relating to monozygotic twins assertedly indicating that their psychological traits are greatly similar. In fact, however, this is not at all the case, since similarities are established only with regard to the most elementary psychophysiological characteristics, and even there serious reservations are stated. Leontiev emphasized that human psychological characteristics develop as a result of an assimilation of the achievements of mankind's socio-historical development, transmitted not in the form of a genetic "recording" but in the form of the phenomena of material and spiritual culture. He also opposed a number of other propositions developed by Neifakh, including the idea that the human race's creative potential is determined by an elite possessing a "fortunate combination of genes". Leontiev noted that with all the stated reservations such a way of thinking leads to a revival of false biologism in theories concerning man and sources of social creativity.

Similarly, A. Malinovsky's position also differed from that of Neifakh.[1] In his opinion, even should the creation of genetic replicas be possible, making hereditary "copies" of gifted individuals would not be at all desirable. Observations indicate that the number of gifted persons is in fact much greater than we usually find, but that their potentials are seldom realized even under favourable conditions. The present task is to realize po-

---

[1] See: A. M. Malinovsky, *Biologiya cheloveka* (The Biology of Man), Znaniye, Moscow, 1973.

tentially available talents. It does not follow, of course, that inherited talents are negligible; however, hereditary prerequisites should be distinguished from social factors.

Efroimson expressed support for the idea that the role of genetic factors in such psychological traits of individuals as talent or a propensity for crime is actually much greater than is usually thought. In his opinion, the role of genetics in determining the overall intellectual level of a person is no less important than that of the environment. On other occasions, including round table discussions sponsored by *Voprosy filosofii,* he defended the need for eugenic measures since, in his opinion, the cessation of natural selection exerts a negative influence on the human race's heredity.

Nazarov, a geneticist, observed that Neifakh had considered only one aspect of man's vegetative replication, namely, the artificial reproduction of highly developed individuals. That method, however, could also be used for preventing hereditary diseases. He disagreed with Dubinin's assertion that no accumulation of unfavourable mutations had been observed.

Philosophers A. Shishkin and M. Mamardashvili stressed the negative social and ethical aspects of the problems being considered. Similarly, V. Kudryavtsev, a jurist, expressed a negative attitude to projects for "improving" the human race and also generally objected to the exaggeration of the role of biological and particularly genetic factors in the social behaviour of individuals.[1]

Another participant, Bochkov, first stressed that the struggle for healthy heredity must become a component of human hygiene in the wide sense of that word, rather than be limited to the elimination of mutational factors or to measures aimed at reducing the intensity of spontaneous mutation processes. He noted, however, that in this context some fundamental questions arose, such as: how should such a struggle be waged if one is to avoid damage to man's most intimate component—his heredity? What should be viewed as uniformed interference into human heredity and what should be seen as a catastrophic failure to in-

---

[1] See: N. P. Dubinin, I. I. Karperts, V. N. Kudryavtsev, *Genetika, povedeniye, otvetstvennost* (Genetics, Behaviour, Responsibility), Moscow, 1982.

tervene? Bochkov expressed his disagreement with those participants who, in his opinion, underestimated the important role played by the original material basis of human development and overemphasized the role of education.

On the whole, the views that were developed by geneticists and Marxist philosophers in that discussion and on subsequent occasions rejoin the views that were expressed in discussions among Western scientists.[1]

The world-view and methodological positions of Marxist specialists in genetics and philosophers concerning various projects for "improving" the human race are quite clear. On the whole they reject these proposals, even though they support specific research activities being carried out in human genetics. There are, of course, differences among them with regard to approaches to some specific problems. In socialist society, where actual social control over scientific research generally, and genetic research in particular, is *effected* both by the government and by the public, research activities are carried out on a broad scale in medical genetics, whose aim is to treat and prevent hereditary deseases, and a network of medical and biological consultation centres is developing and experiments are being carried out in genetic engineering. Both the objectives of research and the means for attaining them are profoundly humanistic in nature and are subordinated to socialism's highest principle and main objective, namely, to benefit man and his free and comprehensive development.

The basic coincidence of scientific and social objectives and means under socialism underlies the coincidence of research and ethical norms guiding scientists and results in a dialectical interrelationship between freedom to engage in research and social responsibility. In capitalist society, the latter two elements are dissociated; thus a conflict is created that is painful for scientists

---

[1] See: *Obshchestvo i zdorovye cheloveka* (Society and Human Health), Moscow, 1973; Hans-Martin Dietl, Heinz Gahse, Hans-Georg Kranhold, *Humangenetik in der sozialistischen Gesellschaft: philosophisch-ethische und soziale Probleme,* G. Fischer, Jena, 1977; *From Biology to Biotechnology,* Brno, 1981; Ivan Kalaikov, *Obshchestviat progres i biologiata na cheloveka,* Sofia, 1983; *Dialektika v naukakh o prirode i cheloveke* (Dialectics in Natural and Human Sciences), Ed. by I. Frolov, Moscow, 1983.

and dangerous for society. For scientists in socialist countries, socio-ethical regulators operate not as forms of "external control" but as forms of self-restraint that are inherent in science itself.

An interpretation of socio-ethical regulators based on what has been described in the West as "Lysenkoism" which, it is asserted, is a phenomenon that cannot be avoided under socialism, cannot be accepted as valid. Not only do such phenomena not derive from the nature of socialism but they deeply contradict socialism, and while they did take place at a certain stage of historical development one should not forget that they were criticized and rejected by the Soviet scientists themselves.

Eventually, the problems and contradictions associated with science's development under socialism find rational solutions that serve both scientific cognition and social wellbeing. The coincidence of scientific and social objectives and means opens unlimited possibilities for research, including that in human genetics. These possibilities are realized gradually as both science and socialist society itself develop, as culture progress and self-awareness grows, and as the new man is formed.

The philosophy and ethics of socialist society preclude manipulative approaches to man. It is on such a basis that socialist countries promote international cooperation in medical studies and studies of human biology. Socialist countries participate in numerous international agreements regulating scientific research on human beings. They are actively opposed to the use of research findings for military purposes and seek to prohibit the development of new biological weapons that would be even more sinister than nuclear weapons.

We perceive the increasingly wide movement of scientists throughout the world for more effective socio-ethical control over scientific research and its technological applications as an expression of reason and humaneness. The humanistic regulation of science, towards which science and society are moving because this is a vital necessity, may provide a basis for a new stage in science's development. An awareness that the social responsibility of scientists and freedom of research are not mutually exclusive is penetrating increasingly deeply into modern science.

This produces *practical* possibilities for both dialogue and joint activities of members of the world scientific community. It also opens new *theoretical* prospects which, in particular, find expression in a new field of scientific research that may be defined as the *ethics of science*.

Close attention is being given to many new as well as "eternal" moral and philosophical problems of human existence that lend a new colour to traditional scientific conceptions. Science's humanistic problems have become component elements of a wider socio-cultural approach encompassing the entire spectrum of man's spiritual aspirations. Let us therefore turn to the moral-philosophical problems in order that we may define more clearly the meaning and significance of the new synthesis of man and humanism.

CHAPTER THREE

REAL HUMANISM AS A NEW MORAL PHILOSOPHY OF HUMAN LIFE

> The secret of human existence is that man must not simply live but must discover why he should live.
>
> *Fyodor Dostoyevsky*

> Only that way of life is genuine which is a continuation of past life and serves the good of both current and future life.
>
> *Lev Tolstoy*

> Although he may be dying even his vestiges retain man's victorious efforts on the road to immortality.... He leaves behind him something unique that he creates through words, deeds, thoughts, even greetings, a handshake or only a silent smile.
>
> *Mikhail Prishvin*

The simplicity and wisdom of these words of great thinkers concerning human life, death and immortality reveal their deep perception of the essence of the eternal problems to which we unavoidably turn — sometimes at the level of everyday awareness and sometimes in fundamental reflections of a scientific, philosophical or even religious nature. Throughout history man's thought has repeatedly returned to these problems, but today their socio-ethical and scientific aspects are formulated and approached in new and sometimes unexpected ways.

In what direction are our evaluations of these problems changing, and how will they influence man's prospects and the self-awareness of individuals both today and in the future? Answers to these questions can be obtained by turning to the scientific *philosophy of man,* to *Marxist humanism,* which we will describe as *new,* since it is replacing the old bourgeois humanism, and as *real* because of its practical and effective nature that is based on the experience of real socialism.

Important though *direct* scientific studies of life, death and immortality may be (including projects to prolong human life and restructure man in the future), this does not fully answer the basic question, namely, *for what reason* this is needed, since we consider only man's natural-biological existence. Yet as a social being man extends beyond that kind of existence, and the answer to that question must also extend beyond direct studies of human life and death in their biological meaning. A *moral-philosophical* approach to that question is needed, described by Lev Tolstoy as *wisdom*—an ingredient that he did not find in science in his time.

This is not the place to consider whether or not that great and persistent critic, who never followed commonly accepted stereotypes, was right. I believe that in many respects contemporary science has abandoned the type of narrow thinking that gave Tolstoy grounds for juxtaposing it to wisdom, particularly on questions involving man, his life, death and his future. Nevertheless, many difficult and important questions relating to world views and to socio-ethical and humanistic problems that concern every man still remain without scientific answers. As a result, man is again turning if not to religion then to something like a "scientific search for God", etc. Quite often as modern science reveals new facets of life in the course of research on human biology and genetics and of man's psycho-physiological possibilities it becomes not only a source of knowledge but also a source of faith, disguised in appropriate scientific phraseology. Often modern answers to "eternal" questions contain various historical reconstructions of thought in which the philosophical wisdom of past ages becomes self-serving as new "prophets" are found who had allegedly already solved all the mysteries of our present and future existence.

This is why it is so important to show the significance of *modern* scientific and philosophical solutions to the problem of life, death and immortality in terms of *new* and real Marxist humanism. This will permit us to approach more closely the solution to the problem of the synthesis of man, science and humanism as it appears from a Marxist point of view. This is not a simple task; aware of this, the author proposes only some *essays* that outline the overall problem in a fragmentary way, and offers a critique of the existing studies in this field.

To begin with, a few preliminary observations would be in order concerning the point that Marxist humanism is both *new* and *real* and concerning its distinction from traditional bourgeois-reformist and utopian forms of humanism that claim to correspond to the current stage in the development of civilization. The intensification of global problems, the course of world events, the development of production and culture impose new demands on man, his activities, his awareness and self-awareness and his will and morality—while the danger of a world thermonuclear war threatens the very existence of the human race. In such a context conceptions have been to the effect that there is a need for an integral world consciousness and a corresponding world wide philosophy (Julian Huxley), as well as for global growth and a new humanism (Aurelio Peccei). It is alleged, moreover, that with regard to such issues Marxism has become hopelessly obsolete and that it should be either abandoned altogether or else reformed in accordance with the ideal of a humanistic socialism, wherein a "new humanism" will become the basis of a utopian "new society" proclaimed, in particular, by Fromm who sought to combine a philosophical-anthropological approach with culturological and socio-political approaches.

Its opponents and reformist bourgeois "Marxologists" have always either sought to represent Marxism as a strictly "economic teaching" devoid of independent philosophical foundations, to juxtapose it to its inherent practical revolutionary content (particularly with regard to the Leninist stage of Marxism) or else to "discover" Marx as a "utopian prophet" who, allegedly, had addressed the problems of man and humanism in his early works and then assertedly reversed himself and developed a teaching from which man simply "vanished", to be replaced by a rigorous "economic determinism" and an absolutization of the "socium". It is true that at various stages of Marxism's, historical development numerous genuine proponents of Marxism, for a number of reasons, occasionally interpreted it in such a way that only one of its facets would appear, thus distorting its nature as an *integral system*. In addition, there have been and there continue to exist objective historical conditions in accordance with which, as Lenin observed, particular aspects of Marxism came promoted to the foreground. At the same time, as a dynamically developing theoretical system Marxism continues to depend, in its interpretation, on the subjective possibilities of those for whom it is a subject of scientific research. In turn, the logic of such research is governed by the particular stage and the circumstances of ideological struggle.

This point was noted by Engels in his comment concerning the materialist interpretation of history: "The *ultimately* determining factor in history is the production and reproduction of real life. Neither Marx nor I have ever asserted more than this. Hence if somebody twists this into saying that the economic factor is the *only* determining one, he transforms that proposition into a meaningless, abstract, absurd phrase ... Marx and I are ourselves partly to blame for the fact that the younger people sometimes lay more stress on the economic side than is due to it. We had to emphasize the main principle *vis-à-vis* our adversaries, who denied it, and we had not always the time, the place or the opportunity to give their due to the other factors involved in the interaction."[1] Of course, all this should be distinguished from conscious and deliberate attempts to distort the *essence* of Marxism as an integral system through "interpretations" that are based on premises alien to Marxism and that either ascribe these premises to Marxism or else claim to "supplement" it.

Marxism is, above all, a *science of man's emancipation and development,* that is, an embodiment of humanism. All its other aspects, including the economic one, important though they may be, are merely a *means* for achieving that objective. And only the emancipation and development of man's nature is an *end in itself.*

This was Marx's view which guided his scientific and revolutionary activities which were concerned with a search for *ways and methods* for achieving man's emancipation and development. Marx found them in a transition to Communism as a new civilization, a *genuinely human society* in which man's *free and comprehensive development* will become possible. This process involves the assertion of a genuine humanism that is real, active and new, in the sense that it replaces the old utopian bourgeois humanism.

Of course, when we turn to an ideal that was discovered nearly a century and a half ago it may seem strange to call

---

[1] "Engels to Joseph Bloch in Konigsberg, September 21-22, 1890", Marx, Engels, *Selected Correspondence,* pp. 394, 396.

it new. A specific characteristic of Marxism's humanism, however, is that it is alive today and above all is turned to the future. It is the human race's true objective and ideal.

While Marxism's new humanism is class-oriented in its origins and is aimed at the working class as the major social force that is able to liberate both itself and the entire human race, it also incorporates organically all the *general human* striving and ideals of the preceding ages. This was generalized remarkably well and expressed laconically by Marx as he defined the "development of the richness of human nature as an end in itself".[1] And this has become the highest objective of the socialist world which embodies the new humanism's lofty ideals in a new civilization. As a step in a new civilization socialism offers practical solutions to man's fundamental problems and makes possible the development of the new man whose qualities correspond to the new conditions of civilization. This is why it is not possible today, under conditions of intensive scientific and technical progress and of intensification of global problems, to think and act in a new way in isolation from the reality of world socialism. And we would leave the firm grounds of reality and hence of new humanism as well, if we were to limit its sphere of action to its own specific characteristics and fail to emphasise the feature that makes it most relevant in the new conditions, namely, its relevance to the entire human race. The universal character of the new humanism and its relevance to the entire human race derives from the latter's own international character and from the world-wide solidarity and cooperation of all working people in their struggle for a future without wars and oppression, for social justice and freedom for the sake of man's development and a flowering of his abilities and gifts.

This is the basis on which a new scientific moral philosophical vision of man is being created through science and the answer that it offers to "eternal" questions of human existence and especially the question concerning the *meaning of life* without an answer to which all other questions lose their meaning.

---

[1] Karl Marx, *Theories of Surplus-Value,* Part II, pp. 117-18.

1. Why Live? The Biological, Social and Moral-Humanistic Meaning of Human Life; the Evolution of Its Length, and Approaches to Its Artificial Prolongation

The Marxist solution to the meaning of human life rests on a comprehensive study of the scientific, social and moral-humanistic problems characterising different yet internally integrated aspects of man's biological and social existence and his life activities and development as an individual and a person in the context of life's general evolution and the assertion of reason on Earth and in the Universe. An emphasis on any particular aspect of man's existence produces, as we have seen, a one-sided and, hence, false interpretation of its essence, and therefore to a one-sided and false idea of the meaning of human life, beginning from the idealistic affirmation of some kind of "absolute principles" of being and of its "pre-established harmony" and ending with a complete denial that the life of man and humanity has any meaning and the assertion that life is a totally accidental and completely senseless phenomenon.

In opposing such views the Marxist conception of the meaning of life derives above all from its inherent value and its nature as an end-in-itself. But it does not limit itself to this and does not ignore the question "why live?". It has already been noted that the Marxist conception organically combines an objective, scientific interpretation of life and of man with a value-oriented approach that also rests on objective, scientific grounds. Human life turns out to be neither accidental (as it may appear to individuals) nor senseless, since individuals and personalities are viewed not simply as such but as parts of the whole—human society. Already in his early works Marx had noted that "the individual *is the social being*. His manifestations of life—even if they may not appear in the direct form of *communal* manifestations of life carried out in association with others—*are* therefore an expression and confirmation of *social life*".[1]

According to Lenin, "this notion (-man) is the *impulse* to realise itself, to give itself objectivity in the objective world

---

[1] Karl Marx, "Economic and Philosophic Manuscripts of 1844", *op. cit.*, p. 299.

through itself, and to realise (fulfill) itself."[1] Yet man differs from all other living creatures above all in that in the course of his individual life he never achieves the "aims" of his historical species life. In that sense he is a continually unrealized adequate being. He is not satisfied with a situation in which, in Marx's words "life itself appears only as a *means to life*".[2] That dissatisfaction and incompleteness is a stimulating cause for a creative activity that is distinct from its direct motivations (material and others). Accordingly, the calling, mission, and task of any person is to comprehensively develop all his capacities.

That is the meaning of an individual person's life which he realizes through society. But that is also, in principle, the meaning of a society's life and that of the human race as a whole, which they realize in forms that are historically diverse, depending on the direct motivating objectives they set themselves and on whether they correspond to those of man's own development. Of interest from this point of view is Marx's note that "Ricardo... wants *production for the sake of production* and this with *good reason*. To assert, as sentimental opponents of Ricardo's did, that production as such is not the object, is to forget that production for its own sake means nothing but the development of human productive forces, in other words the *development of the richness of human nature as an end in itself*. To oppose the welfare of the individual to this end, as Sismondi does, is to assert that the development of the species must be *arrested* in order to safeguard the welfare of the individual... (Sismondi is only right as against the economists who *conceal* or deny this contradiction). Apart from the barrenness of such edifying reflections, they reveal a failure to understand the fact that, although at first the development of the capacities of the *human* species takes place at the cost of the majority of human individuals and even classes, in the end it breaks through this contradiction and coincides with the development of the individual; the higher development of individuality is thus only achieved by a historical process during which individuals are sacrificed, for the interests of the species in the human kingdom,

[1] V. I. Lenin "Conspectus of Hegel's Book *The Science of Logic*", *Collected Works*, Vol. 38, p. 212.

[2] Karl Marx, "Economic and Philosophical Manuscripts of 1844, *op. cit.*, p. 276.

as in the animal and plant kingdoms, always assert themselves at the cost of the interests of individuals, because these interests of the species coincide with the *interests of certain individuals,* and it is this coincidence which constitutes the strength of these privileged individuals."[1] Marx emphasizes that this reflects Ricardo's scientific integrity, which leads him to establish both the coincidence of the bourgeoisie's interests with those of achieving a further development of the productivity of human labour and their contradictions with such a development.

It is also scientific integrity that led Marx himself to recognize the need for establishing the basis of a social ideal in which the development of each person is also the condition for the development of all, and in which social objectives thus coincide with those of the individual. It is precisely the extent to which personal and social objectives coincide—a measure that changes at various stages of history and in various socio-economic formations—that defines the value of human life. That measure dialectically combines the objectives and meaning of the life of the individual and that of society, and these may be conflicting in socio-economic formations based on private ownership and increasingly coincide as social development approaches communism. The movement towards communism is a continual change in the measure of the personal and the social; it is both an increasingly pronounced individualization of the personality and at the same time its integration with society and with the objectives and interests of its development. It is, accordingly, a continual orientation towards the future that lends meaning and value to human life at both the individual and the social levels.

Such an interpretation of the meaning and value of human life is based on the theory of man's social essence. In it, accordingly, social factors do not simply mediate and transform biological factors, but largely "overcome" them. All attempts to infer the meaning of human life from the biological sphere, important though it is in the vital activities of individuals at the level of instincts, subconsciousness, etc., are mistaken as concerns the behaviour of individual persons, which is governed by social factors. The meaning of human life, too, which is determined by these factors, acquires the regulatory function whose role is

[1] Karl Marx, *Theories of Surplus-Value,* Part II, pp. 117-118.

increasingly decisive as the human race's individual and social development proceeds.

This point was well stated by Lev Tolstoy: "Man can view himself as an animal among animals living in terms of everyday preoccupations; he may view himself as a member of a family, and as a member of a society, of a people, living through the centuries, and may and even must view himself (because his mind drives him to this irresistibly) as a part of the entire infinite world living an infinite time. And this is why a thinking man had to do and has always done with regard to infinitely small events in his life capable of influencing his behaviour what is known in mathematics as integration, i.e. define, in addition to his relationship to phenomena of life that are closest to him, his relationship to the entire temporally and spatially infinite world, viewing it as a single whole."[1] In defining the meaning of that "relationship to the whole", Tolstoy considered that it is precisely from that source that man infers a "guidance for his acts". "There has always been only one knowledge that individuals require, namely, a knowledge of one's mission in the particular situation in which man finds himself in this world as well as of the action or else refraining from action that an understanding of that mission implies."[2] For Tolstoy "that knowledge has universally been perceived as the most important knowledge, and has been viewed with the greatest respect. Most often it has been called religion but sometimes wisdom".[3] He considered that "religion is simply an answer to the question concerning the meaning of life".[4]

Of course, one should reject that which Tolstoy ascribed to religion—even though it is true that he interpreted it in such a way that this resulted in his excommunication from the Christian church ... but still it is Tolstoy himself who continues to induce individual Russians, and others as well, to pose and seek to resolve "that question which cannot be resolved by the mind, and that brought me to despair, to suicide, namely, what

[1] L. N. Tolstoy, *Polnoye sobraniye sochinenii* (Collected Works), Vol. 35, p. 161.

[2] *Ibid.*, Vol. 40, p. 427.

[3] *Ibid.*

[4] L. N. Tolstoy, *O Smysle zhizni* (The Meaning of Life), St. Petersburg, 1906, p. 5.

is the meaning of my life?... The answer must be not only rational but also true, i.e., an answer in which I could believe with all my soul, and believe in it unavoidably, just as I cannot avoid believing in the existence of infinity."[1]

Tolstoy saw life not in "playing the role of Solomon or Schopenhauer—that is knowing that life is a silly prank and still continuing to live, wash, dress, eat and speak and even write books. For me this would be repulsive..."[2] He could not admit the "senselessness of life" as do such "negative philosophers" as Schopenhauer and Hartmann, just as he could not see its meaning in personal well-being alone, when "man lives and acts in order to serve his own well-being, in order that all persons and even all creatures live and act only for his own good".[3] For Tolstoy this was "an animal life" not governed by the imperatives of the mind which speaks of the norms of a *general well-being*.

Nor was Tolstoy satisfied by a passive acceptance of what most educated persons mean by "progress". For him to actually live in accordance with progress may be compared to the situation of a man in a boat searching for a direction to follow rather than letting himself be carried by the stream. In lending to his answer concerning the meaning of life religious, Christian forms, Tolstoy proclaimed *love* as a "striving for the well-being of what is outside man, and which remains within man after one discards animal forms of well-being".[4] Love is "life itself; but not a mindless life that is perishing, but a life that is serene and infinite ... It is life's most joyful activity".[5] The "movement of common life", moreover, is leading increasingly towards a harmony and unity of individuals produced by subordination to reason. That harmony, in his opinion, is the highest good—"a perfect union in a perfect, highest form of reason".[6]

Placing an emphasis on the ideal aspects of man's world and essence, his thinking, reason and morality ultimately leads one

---

[1] L. N. Tolstoy, *Polnoye sobraniye sochinenii* (Complete Collected Works), Vol. 24, p. 14.

[2] *Ibid.*, Vol. 23, p. 28.

[3] *Ibid.*, Vol. 26, p. 370.

[4] *Ibid.*, Vol. 23, p. 8.

[5] *Ibid.*, Vol. 26, p. 394.

[6] *Ibid.*, Vol. 35, p. 177.

to regard the problem concerning the meaning of human existence as essentially unnecessary since it is already embodied in the very definition of man. That thought has long been recognized in philosophy. Let us recall Kant's words to the effect that "one can no longer ask of man ... as a moral being for what reason (quem in finem) he exists. His very existence contains the highest objective to which he can subordinate all of nature within limits of his abilities."[1] But then such a calm and truly philosophical observation (in the spirit of German philosophy) is not given to every thinking person without effort. Each individual seeks *his own* answer first to the problem of the meaning of his *own* life and only then of his *species'* life. That search can be highly emotional, tormenting and with tragic overtones as was the case, in particular, in Russian history, where in the works of such authors as Lev Tolstoy, Fyodor Dostoyevsky and Vladimir Soloviev philosophical thinking acquired highly specific forms.

The search for an answer to the question concerning the meaning of life has often led these thinkers to errors and extreme positions. But one should recall that in the extremely difficult social conditions under which the search was taking place, its very fact constituted an act of daring, of suffering for the common good, for everything that is *truly human in man* that Tolstoy described as "infinitely great and undefinable". We may disagree, for example, with Tolstoy's or Dostoyevsky's interpretations of the ideal defining the meaning of life, but the fact that they emphasised the active moral principle in man can be accepted by us even today.

This is why there is such a contemporary ring to much of what Dostoyevsky described contemptuously as "the philosophy of the environment". He considered that "...there is only error and deceit—a deception that is rather tempting", for by making man "dependent on each error in social arrangements the theory of the environment leads man to complete loss of identity, to his complete liberation from any personal duty, from all autonomy, to the most repulsive slavery that one can imagine".[2]

[1] Immanuel Kant, *Kritik der Urteilskraft*. Ed. by Karl Vorländer, Verlag von Felix Meiner, Leipzig, p. 305.

[2] F. M. Dostoyevsky, *Polnoye sobraniye sochinenii v tridtsati tomakh* (Collected Works in thirty volumes), Vol. 21, pp. 16, 17.

And it is only natural that this fundamental question—"why live?" is raised so acutely in Dostoyevsky's works and is addressed directly to man's conscience, particularly in the context of "refashioning of mankind after some new pattern"[1] in accordance, in particular, with conceptions of socialism and anarchism as he perceived them and as they occasionally manifested themselves in real life. Dostoyevsky explored life at those "points of fracture" that bear its very essence, as it were. This is why he posed so sharply the question "what is all this for?" "Surely I haven't gone through suffering so that some future harmony should be manured, for somebody else, with me, my evil deeds and my sufferings .. and therefore renounce the higher harmony completely. That harmony is not worth a single tear of the tormented child that beat his breast with his little fist and with unexpiated tears, prayed in stinking privy to his 'dear kind God'. ... I want no harmony; out of my love for mankind I do not want it ... even if I'm in the wrong."[2]

Such an obstinate rejection of future well-being, if the road to it and the means to achieve it may contradict morality and humaneness, reveals new facets in interpreting the meaning of human life. And it does not require much imagination to correlate all this, from the positions of Marxism and communism, with our current concerns and reflections as regards not only man's and the human race's past but also their future. Is this not the reason why we turn so often today to the ideas produced by our great visionaries in the fragile and quivering world of moral ideas. What Dostoyevsky said about them in his famous speech on Pushkin also applies to them—they stand before the future as an injunction and a prophecy. Their influence on the present generations is precisely *moral* and this is why they are true creators of the foundations of *moral philosophy* in Russia.

Of course, as noted earlier, it is very harmful and fruitless to seek to modernize the past and to galvanize in the history of Russian culture the elements that did not and do not serve progress, counterpose the national to the international, sever Russian tradition from world culture and, above all, absolutize that tradition and lead it away from the high road laid here

[1] Fyodor Dostoyevsky, *The Karamazov Brothers,* Progress Publishers, Moscow, 1981, p. 356.

[2] *Ibid.,* pp. 373, 374, 375.

by Marxism, Lenin and his disciples and followers. This, I am sure, especially applies to the views of several of Russian thinkers who turned precisely to the moral-philosophical problems of the meaning of human life, death and immortality. What can draw our attention and be used here and what should be rejected and left in the past, "leaving the dead to bury the dead"?

In the history of Russian culture the moral-philosophical problematique of life, death and immortality contained not only specific but also general features that link Russian culture with world culture. Moreover, one should not overlook the eternal and unseverable link between the problem of the meaning of human life and the attempts to unravel the mysteries of non-existence and death, and also man's desire to live eternally, either materially or at least spiritually, and to morally vanquish death. This is aptly expressed by Anna Akhmatova:

Will you not repeat once more
The word that has vanquished death,
And the answer to the riddle of my life?

No one has yet said and I think no one ever will say such a "word", about which this remarkable Russian poetess dreamed, as she reflected deeply on life's "eternal" questions, a "word" that could both vanquish death and unravel the riddle of life. And can there be such a "word" if it must be both yours and mine. I do not wish to say that there cannot be philosophical solutions of general significance in this regard on the basis of findings of the socio-biological sciences, but this will depend entirely on the extent to which natural scientific approaches to these "eternal" questions can be combined organically with moral-philosophical approaches. The specific forms of such a combination (or severance) determine the peculiarity of this or that cultural tradition, if not culture generally. They constitute its humanistic "core" which manifests itself differently in the history of each people and of the entire human race.

An interesting tradition of moral-philosophical reflections concerning man's death and immortality has developed in the history of Russian science and culture. It is found not only in the works of major literary figures and scientists but also in their diary notes, which in most cases were not meant for publication. The intimately emotional and personal aspect of these problems is clearly evident in these notes. This is especially true of Tolstoy's diaries. But our other great explorer of life's secrets, Dostoyevsky, also discussed these problems at length not only in his famous novels but in his letters and in his *Diary of a Writer* as well. It is true that in different measure and for different philosophical reasons both entered into contradictions with the scientific knowledge of their day at least as they interpreted it, for reasons that were not always well grounded. Beyond this they linked their discussions to "the basic and and noblest" idea of human existence, namely, the need, and in their view the inevitability of a belief in the immortality of the human soul.

In his *Diary of a Writer* Dostoyevsky stressed that the "highest word and highest thought" never "die and never vanish completely, they can never vanish, if only they are expressed but once—and this is a remarkable characteristic of mankind".[1] For Dostoyevsky, like the idea of love for mankind in general, "the highest idea of human existence" "is one of the human mind's most unfathomable ideas". In his view, as a highly moral idea it governs human life and makes it possible to solve "the highest, the most primary questions: 'why live'... 'what keeps on earth' individuals of the highest types who do not live in terms of material interest alone". The moral idea also leads man towards personal "absolute self-perfection in the future, in the ideal", for according to Dostoyevsky, "one cannot become a man all of a sudden, one must form oneself into a man ... and even if that ideal were possible, incompletely formed persons would not be governed by any rules, even the most obvious ones".[2]

It is interesting, however, that this theme, which Dostoyevsky resolved on a religious-idealistic basis unacceptable to us, also rings throughout his novels, and, moreover, in non-religious variants as well. We find this not only in *The Karamazov Brothers* but also earlier, in *A Raw Youth,* in particular, Versilov's monologue that describes a state of society in which the people have become disappointed with the idea of the "immortality

---

[1] F. M. Dostoyevsky, *Collected Works,* Vol. 24, p. 47.

[2] *Ibid.,* Vol. 25, p. 47.

of the soul".[1] Of course, one can always say that these are not the words of Dostoyevsky himself but those of his hero, but then how little understanding there would be of Dostoyevsky if one were always to hide behind that phrase. While I am not a specialist in literature I have studied the moral-philosophical aspects of Dostoyevsky's creative works for many years and I have long come to know the importance of being careful in this respect. It is, of course, clear that Dostoyevsky's works abound in contradictions. But what is important is his own emphasis on the immense significance of the "moral idea", rather that the specific way in which it was expressed. It is, of course, absurd to modernize Dostoyevsky's views and seek, as certain authors

---

[1] Since one of my well-intentioned opponents cites this as a proof of Dostoyevsky's hostility to such a state of "orphanage", I wish to cite that monologue in full to support my conclusion. "Having become orphans, individuals immediately began to huddle against each other more closely and with greater affection; they were ready to grasp each other's hands understanding that now they alone are everything for each other. The noble idea of immortality would vanish and it would be necessary to replace it; and the entire vast overflow of former love towards Him, who was immortality, would turn towards Nature, the world, towards people and every little thing. They would come to love the earth and life boundlessly, and this to the extent to which they would gradually become aware of their own transitoriness and finiteness and, moreover, with a new different love. They would begin to note and discover in Nature phenomena and secrets whose existence they did not suspect in the past, for they would look at Nature with new eyes as lovers look at their beloved. They would awaken and hurry to kiss each other, hurry to love, realizing that their days are short and that this is all that they have left. The would work for each other and each would give everything he has to all, and this alone would make him happy. Every child would know and feel that every human being on earth is like father and mother to him. Looking at the setting sun each person would think: 'let tomorrow be my last day, but this does not matter, I will die and they will all remain, and their children after them,' and the idea that they will remain, continuing to love each other and tremble for each other, would replace the thought of meeting beyond the grave. How they would hurry to love in order to quench the immense sadness in their hearts. They would be proud and bold with regard to themselves but would be timid for each of her; each would tremble for the life and happiness of every other. They would come to need each other and would not be ashamed of this as they are today, and would caress each other like children. When meeting they would gaze at each other with a deep and meaningful look, and in their look there would be love and sadness..." (*Ibid.*, Vol. 13, pp. 378-379).

are doing, to "incorporate them" into Marxism—on the grounds that they are so close to a *genuine* communism as opposed to "the barracks communism" against which they are said to be solely addressed. But it is also unwise to underestimate the progressive moral value which many of Dostoyevsky's ideas have and which today help Marxists and Communists not only in opposing bourgeois lack of morality, cynicism and lack of spiritual content, but also in developing new relations among persons in accordance with "mankind's loftiest ideas". In any event we cannot and must not surrender him to our opponents, who seek to set Dostoyevsky's humanistic ideas against Marxist, real humanism.

We know that in *The Possessed,* for example, Dostoyevsky's humanistic position was aimed at those forms of "revolutionary action" that have since been called "Nechayevshchina". And while they do not have anything in common with the Marxist theory of revolution (and its teachings concerning the interrelationships of the individual and society) and concerning the objectives and means for achieving individual and social well-being, it is precisely such an approach that continues to be ascribed to Marxism today together with a sharp rejection of the "Karamazov questions" concerning a price for a future "higher harmony" that is unacceptable to a humanist.[1]

It is immoral to describe in this way a teaching that has proclaimed the absolute revelation of creative gifts, and integral character of human development as the end-in-itself of historical development. One also cannot fail to recall the lofty moral criteria to which the founders of Marxist theory adhered in their own lives. It was clear to Marx already in his youth that if a man "works only for himself, he may perhaps become a famous man of learning, a great sage, an excellent poet, but he can never be a perfect, truly great man. "History calls those men the greatest who have ennobled themselves by working for

[1] This was initiated by R. V. Ivanov-Razumnik, who in criticizing the "positive theory of progress" asserted that for Marxists the "Karamazov questions" are solved in such a way that through our suffering and sacrificing of our lives for the future of mankind we obtain the meaning, the objective, and the justification of both individual human life and worldwide human history (R. V. Ivanov-Razumnik, *O smysle zhizni* /Concerning the Meaning of Life/, Moscow, 1908).

the common good, experience acclaims as happiest the man who made the greatest number of people happy."[1] It is this that made it possible for Marx to persist in his efforts to his very last hours, sacrificing his health and happiness of life. But he also wrote together with Engels that "if correctly understood interest is the principle of all morality, man's private interest must be made to coincide with the interest of humanity".[2] And this, as Engels stressed, would happen "in a society in which community of interests has become the basic principle, and in which the public interest is no longer distinct from that of each individual!"[3]

This is the core of the Marxist solution to the problem of the individual's relation to society and the basis for understanding the meaning of individual human lives. Here we see the Marxist position expressed not only in terms of theory, but also in intimately emotional and personal aspects. The very mentioning of these aspects, I believe, invalidates Marxism's enemies' gross slander to the effect that it is "soulless" and that it ignores everything personal, including the moral-philosophical problems of the meaning of life. It also invalidates continuing attempts to solve these problems somewhere near or apart from Marxism—either in existentialism or in appropriately refurbished idealistic philosophical theories, including, in the case of Russian history, the views of Tolstoy, Dostoyevsky and Soloviev.

It would be senseless to borrow something directly from Tolstoy or Dostoyevsky. For Marxism possesses its *own tradition* in considering moral-philosophical problems from a position of real humanism which is, moreover, developing in *practice*. But we by no means wish to say that all these problems are already solved, that there are no theoretical and practical contradictions and, hence, no further prospects for moral-philosophical research, and, specifically, concerning the problem of the meaning of life. On the contrary, the development of both man and of society, of science and of culture in modern conditions turns our atten-

---

[1] Karl Marx, "Reflections of a Young Man on the Choice of a Profession", in: Karl Marx, Frederick Engels, *Collected Works,* Vol. 1, 1976, p. 8.

[2] Karl Marx and Frederick Engels, "The Holy Family", *op cit.*, pp. 130-131.

[3] Frederick Engels "Speeches in Elberfeld", *op. cit.*, p. 249.

tion to the new aspects of these problems. The threat of a nuclear catastrophe, the growing intensity of global problems, the development of microelectronics and robotics, of gene technology, reanimation science and psychophysiology—all this calls forth new aspects of the problem of the meaning of human life. But our new interest in the future also intensifies our interest in the past. And there we may find not only historical data for further reflection but also powerful positive stimuli for analyzing past cultures. Of course, it is also necessary to adhere to the principle of objectivity, avoid modernizing and even more so absolutizing historical experience, and relate it correctly to Marxism's position and traditions. This also relates to the moral-philosophical ideas bearing on the meaning of life, and on human death and immortality that were so vividly expressed in Russian history, in particular by Tolstoy and Dostoyevsky[1].

This is also true of a number of Soloviev's moral-philosophical ideas. There is, of course, an inherent link between many of Soloviev's moral-philosophical ideas and the relevant reflections of both Tolstoy and Dostoyevsky. Accordingly, there, too, a certain *tradition* was created that calls for analysis in order to establish its "rational meaning" which the enemies of Marxism and Leninism frequently misuse today. Because this is so important, Marxist thought cannot aflord, I believe, to forgo such an analysis.

Of course, like Tolstoy and Dostoyevsky, and perhaps to even a greater extent, Soloviev, too, was full of contradictions which, nevertheless, represented, in Lenin's words, "a reflection of the extremely complex, contradictory conditions, social influences and historical traditions which determined the psychology of various classes and various sections of Russian society in the *post*-Reform, but *pre*-revolutionary era."[2]

Soloviev sought to provide a theoretical formalization (on religious-idealistic basis) to a number of ideas concerning the meaning of human life that are consonant with Tolstoy's and

[1] Understandably, Soviet literature in this field has placed a greater emphasis on the links with Marxism of revolutionary-democratic thinking, which gives rise to the study of the problem of the individual and society and of the related ethical and moral-philosophical questions in the works of Georgi Plekhanov and especially V. I. Lenin.

[2] V. I. Lenin, "L. N. Tolstoy", *Collected Works*, Vol. 16, 1977 p. 325.

Dostoyevsky's approaches to moral-philosophical issues, even though they also greatly differ from them in many respects.

In his major work entitled *The Justification of the Good*[1] Soloviev, following the basic principles of the moral philosophy that he was developing, expressed, even though in a religious and mystical form, a number of profound ideas that can (when re-examined, of course) provide us with abundant material for reflections on the meaning of life.[2] For Soloviev "to show that good is truth" means to find the "only correct way in life that is true to itself in everything and to the very end", while specific *moral norms* provide the condition for achieving life's true objective. These must be followed by a man who has consciously set himself this objective.[3]

Defining the moral meaning of life in "its preliminary sense", Soloviev dissociates himself from both the pessimists (theoreticians and "practitioners", i.e., suicides) and from the Nietzschean approach, according to which the meaning of life is to be found in its aesthetic aspect, in that which is strong, majestic, and beautiful and that may be developed further to the point of creating superhuman grandeur and a new beauty of the greatest purity.[4] He does this, however, in a religious-Christian manner, in which he does not view as power what is powerless vis-à-vis death and contrasts the latter to faith in salvation, which was proclaimed by "the first to be reborn from the dead", i.e., by Christ whose resurrection assured "eternal life" to his followers.

Here we have touched on the core of all idealistic teachings about the meaning of human life which are always associated with the postulate of an eternal existence of some suprapersonal forces opposing the individual's death in his corporeal being. But materialist conceptions, too, cannot escape the need for

---

[1] See *Sobraniye sochinenii Vladimira Sergeyevicha Solovieva* (Collected Works of Vladimir Sergeyevich Soloviev), 2nd edition, Vol. 8, St. Petersburg.

[2] While considering the views of Soloviev relating to the meaning of life I do not claim to present a comprehensive characterization of his general philosophical and ethical views.

[3] V. S. Soloviev, *op. cit.*, Vol. 8, p. 3.

[4] The ideological opposition of Soloviev to Nietzsche can also be seen in the titles of their respective major works—*The Justification of the Good* (Soloviev) and *Beyond Good and Evil* (Nietzsche).

defining the meaning of life in some way that does not oppose it to the meaning (or lack of meaning?) of death. And later we will see that this is also important for other fundamental moral-philosophical problems.

Soloviev believed that only faith in God affirms the undying power and beauty which are inseparable from good. "We cannot invent a genuinely good universal meaning of life or an internal connection of individual units the great whole, it is given from eternity."[1] However, it must then be defined as something that is *within* rather than *outside* of man. In Soloviev's view "he who wishes to accept the meaning of life as an external authority will end by taking for the meaning of life the meaninglessness of his own arbitrary rule. There must not be an external formal relation between man and that which is the meaning of his life. An external authority is needed as a transitory element but it cannot be eternized and recognized as constant and final norms. The human *Self* can only be widened through man's inner heart-felt reciprocity with what is larger than himself, and not through merely a formal subordination to it, which in effect changes nothing."[2]

This is just as true as the proposition that "life's positive meaning" cannot be inferred from the subject alone and from the internal moral actions and states of one individual. In criticizing the latter as an *amorphism* or *subjectivism,* Soloviev refers to the falseness of the teaching—probably Tolstoy's—to the effect that Nature itself endows everyone with inner meaning and inner good, but that these are suppressed, distorted and transformed into something senseless and evil by the state, church, and culture generally. Here again as a "counter-argument" there is a reference to death, which does not vanish following the demise of social institutions and yet is incompatible with "life's positive meaning" which is said to "appear and realize itself spontaneously among a formless mass of 'wandering' truthseekers."[3]

Soloviev considers that these two opposite errors, namely, self-abnegation of the personality when confronting historical forms of life are accepted as external authorities and its self-assertion in opposing these authorities—are overcome by defining good as

[1] Soloviev, *op. cit.,* Vol. 8, p. 15.

[2] *Ibid.,* p. 17.

[3] *Ibid.,* p. 18.

an absolute ("unconditional") moral value, noting that "in principle or by his destiny man is an *unconditional* inner form for good as an unconditional content; everything else is conditional and relative. Good by itself is not subject to any conditions, it itself conditions everything else and is realized through everything. The fact that it is not conditioned by anything constitutes its *purity;* the fact that it itself conditions everything constitutes its *fullness;* and the fact that it is realized through everything constitutes its *force* or effectiveness."[1]

This is why Soloviev believed that a person's task is determined by the inner properties of good, while the moral meaning of human life lies in serving good—a good that is pure, comprehensive and omnipotent. In that connection he refers to Kant who, he believed, achieved in part the task of creating moral philosophy (a justification of the purity of absolute good) but rejects "dialectical moments of abstract ideas" (as in Hegel) and "empirical complications of natural facts" (as in Herbert Spencer).

It would be interesting and informative to analyze more closely and rethink materialistically many of Soloviev's ideas concerning his moral-philosophical explorations. But that attractive prospect would take us far afield in terms of our discussion's basic objective which concerns the meaning of human life. It should be said that this question constitutes the pivot of Soloviev's moral philosophy and that it is therefore important at least to mention some of his ideas. Above all, this concerns the view that good is an absolute value in the sense that it rejects those forms of relativism that could result in viewing expediency and necessity as a relative good but at least not as partial evil. I believe that this does not contradict the teaching that human race's various aesthetic principles are relative in the sense that they are concrete-historical in nature, while orienting us more closely on certain principles and objectives that are postulated in ideal form that play a regulating role and without which the human life would be unfree and devoid of meaning. These principles and objectives include, in particular, many values of communism that are common to all mankind and that appear as relative today but will be absolute in the future and that make it possible already now to arrive at positive and negative judgements.

[1] *Ibid.*, p. 22.

While Soloviev lends religious-idealistic forms to these ideas he recognizes clearly that "without a pure good and without the possibility of distinguishing good from evil in all practical issues and to state unconditionally *yes* or *no* in every single case—life would be altogether deprived of morality and integrity; without a fullness of good, without the possibility of associating it with actual relations and in all cases to justify good and to remedy everything with good, life would be one-sided and impoverished; finally, without the power of good, without the possibility of its final triumph over everything, to the 'last enemy' including death—life would be barren."[1]

On these grounds Soloviev rejects utilitarianism, even though he does note that in the teachings of its best representatives it addresses itself to ideas, lying far beyond its narrow confines of the "*solidarity* of all of mankind that links the personal well-being of each individual to universal well-being."[2] Soloviev notes that "the essence of morality in itself is *one,* it is the *oneness* of man inherent in his nature as an existing *norm* that is realized in life and in history through man's struggle with everyday life's centrifugal and divisive forces, as a moral *activity*."[3] But according to Soloviev, man achieves that "oneness" through God, and it is there, moreover, that it is transformed from a potential possibility into actuality. A number of ideas follow from this that cannot always be interpreted in a dogmatic-Christian spirit. In particular, this concerns the proposition that "the world process is not only a process of development and further perfection but also one of *gathering the Universe*."[4] Having defined that objective, Soloviev concludes that "the supreme task of man as such (of pure man) and that of the purely human sphere of being is to gather the Universe into an idea; the task of Man-God and of the kingdom of God is to gather the Universe into *reality*."[5]

What is then the final outcome of the moral philosophical

---

[1] Soloviev, *op. cit.*, Vol. 8, p. 23.

[2] *Ibid.*, p. 160.

[3] *Ibid.*, p. 175.

[4] *Ibid.*, p. 220.

[5] *Ibid.* There the connection between Soloviev's ideas and N. F. Fedorov's "philosophy of the common task" is entirely clear. See Nikolai Fedorovich Fedorov, *Sochineniya* (Works), Moscow, 1982.

interpretation of the meaning of human life that is provided by the author of *The Justification of the Good?* We find it in the way in which Soloviev resolves the question concerning the reciprocal relations between individuals and society and between their interests and objectives. In his opinion, "it is actually impossible to oppose the individual and society to each other, it is not possible to ask which one is the object and which only the means."[1] Such an opposition is impossible since "each individual person is only a point of intersection of an infinite multitude of reciprocal relations with another person and with others, and to separate him from these relations means to take away from him real contents of life, to transform a person into an empty possibility of existence."[2] Each individual's personality "is a possibility for achieving an unlimited reality or a special form of an infinite content."[3] In affirming the infinite character of the human personality as an axiom of moral philosophy, Soloviev opposes both individualism and those forms of collectivism that "because they see only social masses in the life of the human race view individuals as negligible and transitory elements of society who do not possess their own rights and who may be ignored for the sake of a so-called general interest."[4]

There, I believe, Soloviev rises to the heights of humanism. In criticizing all forms of an illusory collectivism he touches on the *dialectics* that governs relations between individuals in society, that was then already discovered and scientifically grounded by Marxism but still not known to a Russian thinker who conceived socialism in the spirit of Dostoyevsky's *The Possessed*. "And what kind of society is this," he asked, "that consists of creatures deprived of rights and of personality, of moral naughts? Will this be in any case a *human* society? What will it be and where will it find its dignity and the inner validity of its existence, and on what *force* will it rest? Is it not clear that this is a sorry chimera that is as unfeasible as it is undesirable? And is not the opposite ideal of a self-contained individual another such chimera? Take away from a *real* human personality everything that in some way derives from its ties to social or collective objectives and

[1] Soloviev, *op. cit.*, Vol. 8, p. 227.

[2] *Ibid.*,

[3] *Ibid.*, pp. 228-229.

[4] *Ibid.*, p. 229.

you have an animal entity representing only a pure possibility or empty form of man, i.e., something that actually does not at all exist."[1]

It is extremely interesting to note that Soloviev perceived the personality as a force "engaging in intelligent cognition and acting morally" possible only in the "form of a social existence". The latter, in his opinion, is inherent in "the very definition of an individual."[2] "The very idea of good or of a moral evaluation is not simply a consequence of social relations, as many believe, but it is only too evident that *realization* of this idea or real development of human morality is possible for the individual only in the social environment and through mutual interactions with it. And in this most important respect society is simply a social realization of an individual's content."[3]

Such a profound understanding of the problem could advance humanistic thought further if it were not, first of all, subsequently immersed in religious speculations, and secondly, if it did not absolutize the moral-philosophical aspects of the existence and development of the individual in society. Soloviev believed that the three basic components of personal-social life, namely, religious, political, and prophetic, operate in history as major steps in human consciousness and the life system, i.e., clans, national states and "world-wide life communication" as the future's ideal. This is the basis on which Soloviev defines the meaning of human life. "By its very nature, both from above and from below, human life is an involuntary participation in the progressive existence of both the human race and the entire world; but its virtue as well as the meaning of the entire Universe require that this involuntary participation of each individual in everything become voluntary, increasingly conscious and free, i.e., truly *personal,* that each person understand increasingly and carry out the *common* task as *his* own."[4]

There can be no objections to such an understanding of the meaning and mission of human life as they appear to *each individual personality*. But in emphasizing the point that man is a *moral* being, Soloviev also largely limits himself to the sphere of

---

[1] Soloviev, *op. cit.,* p. 229.

[2] *Ibid.,* p. 230.

[3] *Ibid.*

[4] *Ibid.,* p. 232.

morality alone. Unquestionably, it is also there that one finds the strong side of Soloviev's humanism, for it directly follows from it that "*no man under any conditions, and for any reason can only be viewed as a means* for achieving any external objectives—he can not be only a means or *instrument for the well-being of another person,* or an entire *class* or, finally, for so-called *common welfare,* i.e., the well-being of the majority of other persons."[1] Further on Soloviev explains that this "common welfare" or "common utility" has the right not to man as a personality but to his activity or work to the extent that while serving the purposes of society it also ensures a worthy existence to him. Common welfare is conditioned by the welfare of individuals as an end in itself, i.e., in essence it can, according to Soloviev, itself become a moral principle when it becomes truly common and concerns not just many or a majority but *all without exception.*

Of course, stating the question in this way is simply an expression of moral maximalism and utopianism, since it does not take account of the *real* movement of history and the unquestionable fact that if not absolutely, then at least relatively "many" and even more so "a majority", when they express progressive needs of social development, also endow society with a moral principle that makes it possible to move forward along the path of good.

At the same time Soloviev's position is in many respects attractive to Marxists and highly relevant today. I have in mind his formulation of the problem concerning the *human relations* under which, for example, social relations in the sphere of material production become moral: first, he insisted that the sphere of economic activity should not be isolated and should not assert itself as independent and self-sufficient; secondly, he insisted that production activities not be effected to the detriment of the human dignity of producers, who would then merely be its "instruments", but give them the material means for a worthy existence and development. It is clear, of course, that it is insufficient in this respect to limit oneself to the sphere of moral requirements, but, on the other hand, they cannot be ignored, for this would contradict the more comprehensive Marxist for-

[1] *Ibid.*, p. 296.

mulation of the problem of man's development as an end-in-itself.

While Marxism has its own scientific tradition in considering these problems, many of the ideas of this Russian thinker could, I believe, stimulate the active constructive work of the mind in the fields relating to the study of man and his future that are unfortunately little developed. Above all this concerns the *moral meaning* of human life. But the method employed in resolving the problem relating to the meaning of human life is also important in other spheres of human activity including the economic sphere and that of man's interrelations with Nature.

The deep linkage that exists between the meaning of human life and the problem of longevity, death and immortality, can be traced throughout the history of both science and philosophy. It was already clearly stated by Seneca who noted that what is important is not whether one lives a long life but whether one lives it correctly. Similarly, Leonardo da Vinci observed that any life that is well-lived is a long life. Montaigne, too, emphasised that point when saying that the measure of life is not its duration but the way you use it. Clearly, in this instance the measure of life is defined by its positive socio-individual and moral form. That is the criterion that can also serve as a point of departure in resolving complex problems relating, in particular, to projects for increasing human longevity.

Biologically, the individual is always in a sense only a *means* for the species as a whole, since it is precisely through the individual's adaptive vital activities culminating in the reproduction of progeny and in death that the species provides for its own existence as a specific form of life possessing different time dimensions than those of individuals. But while in a biological sense Nature becomes "indifferent" to and "loses interest" in individuals after they complete their reproductive function, for the human individual who has become a person it is no longer Nature but society that difines the measure and value of his life. And it is precisely at the point where Nature recedes into the background that the interest of society increases, for the development of the individual constitutes both the end and the means of the existence and development of the human race—of the *Homo sapiens* species—as a social community and as a bearer of reason and culture on Earth.

This dialectical contradiction serves as the basis and motive force in the evolution of human longevity. And it is strongly felt in the life of each personality who, having become a personality, does not cease to be an individual and is, of course, not "liberated" from the natural-biological conditions and factors of his existence. It is also important to note that even in a biological sense a relative increase in an individual's longevity may be associated in evolutionary terms with a greater stability and resilience of a progressing species.

Individual longevity in itself, however, is not an absolute criterion of the stability and reliability of living systems, a condition for the length of the viability of a species in particular environments. That duration or measure of life of a species finds expression in the replacement of generations which is accompanied by the aging and death of individuals. Moreover, as noted by I. I. Shmalgauzen, a Soviet scientist, "among animals increasingly perfected resilient forms of individuality develop . . . Animals possesing a fully resilient active individuality proliferate only through selected sex cells. Sooner or later the entire remaining body perishes. Death is a price for acquiring higher individuality and prolonged existence. . . Thus the existence of individuals is always limited in time. If not death then other causes result in a loss of individuality."[1]

By highly developed individuality Shmalgauzen meant "the form that is characterized by the greatest activity."[2] It is also legitimate to extend these characteristics to man as an individual representing the species of *Homo sapiens,* although on a subjective plane these propositions bring to mind reffections and conclusions that are rather sad. Yet they to some extent echo Marx's words noted earlier concerning the "price" that is paid for a higher development of individuality and the point that in the world of human beings as in the world of animals and plants the interest of the species always prevail over those of individuals, since the interests of the species coincide with the interests of particular individuals. At that point, however, the analogy ends, since in the presence of specific social conditions

[1] I. I. Shmalgauzen, *Problema smerti i bessmertiya* (The Problem of Death and Immortality), Moscow-Leningrad, 1926, p. 80.

[2] *Ibid.*, p. 79.

(communism) the development of individuals as persons coincides with the "interests of the species".

The problem of human longevity, aging and death of individuals and personalities which has long preoccupied human thought continues to remain one of the greatest riddles. An enormous bibliography exists of studies (more than 60,000) of various aspects (biological, demographic, gerontological, psychological) of the problems of man's aging and death in their interconnection and mutual interdependence. Today, moreover, that problem's social, philosophical and moral-ethical aspects are becoming especially important.

A substantial aging of the world's population is currently taking place.[1] It is anticipated that by the year 2000 the number of elderly persons will increase to 600 million. This urgently raises the issue of creating conditions that will make it possible for them to prolong their active life and continue to participate in labour and social activities. This has found reflection, in particular, in the Guidelines for the Economic and Social Development of the USSR for 1981-1985 and for the period ending in 1990, in which the problem is explicitly formulated to "implement a system of measures to increase people's life span and labour activity, and build up their health."[2] In the USSR this is also the aim of a comprehensive scientific programme on "Prolonging Life", in which dozens of research institutes and organizations participate under the coordination of the Academy of Medical Sciences' Institute of Gerontology. The motto "A Full Life In Old Age" is guiding a number of lines of activity among scientists and medical workers throughout the world and, in particular, of activities sponsored by the World Health Organization (WHO).[3]

Naturally, its implementation is primarily determined by socio-

---

[1] In the classification adopted by the World Health Organization (WHO) "elderly persons" range from 61 to 75 years for men and 55 to 75 years for women. This is followed by old age.

[2] *Documents and Resolutions. The 26th Congress of the Communist Party of the Soviet Union,* Novosti Press Agency Publishing House, Moscow, 1981, p. 164.

[3] The United Nations World Assembly on Aging (Vienna, 1982) has developed an international plan of action aimed at increasing the longevity of aged persons. That plan was approved by representatives from approximately 120 countries.

economic factors, and there socialism possesses unquestionable advantages, for it makes possible a practical realization of present and future humanistic norms. However, there are also, problems of a global character in which a considerable role is played by philosophical (socio-ethical and moral-humanistic) problems that closely depend on the solution to the question of the meaning and value of human life. Their correct formulation and solution require a scientific understanding of the essence of the processes associated with the life, aging and death of human beings which must be viewed in the context of interaction among biological and social factors that determine the measure of life and of the biological and social duration of individuals' vital activities. Marx observed that while during his life a person passes through a number of ages all human ages also coexist, distributed among different individuals.[1]

This exerts a substantial influence on the *human* perception of the duration of time and of the measure of life and also makes it possible to perceive the dialectical unity between individual time and historical time, biological time and social time.

This is why when we consider the possibility of increasing longevity a fundamental issue arises, namely, what particular time do we have in mind and wish to prolong: the individual's time or his socio-personal time, which, in spite of its objective character, is highly subjective and is measured in terms of very different, emotionally-laden factors.[2]

The question *why* the individual and society require a prolongation of human life may not even arise if life itself is taken as an absolute and self-sufficient value. That same question may also be "solved" spontaneously in the course of the historical process, and in particular, may take the form of greater longevity. That this is in fact how it is being "solved" is shown by the corresponding historical and demographic data.

These has been an evolution in human longevity throughout the human race's history and it continues to this day. What is meant here is a social life span in which the environment and conditions of life play a determining role, altering substantially

---

[1] K. Marx, F. Engels, *Sochineniya* (Works), Politizdat, Moscow, Vol. 46, Part II, p. 141.

[2] Let us recall Anna Akhmatova's poetic exclamation: ". . .What we shall to do with the horror that was once named the race of time?"

the operation of biological factors.[1] While demographic data are not always accurate and differ in the works of various authors, they do express general principles and tendencies relatively comprehensively as well as their specific distribution in today's developed and developing countries. A consideration of social factors makes it possible to subdivide the process of social aging of man into forms that are normal, when it occurs naturally as reserves of the organism are depleted, and pathological, in which one notes the negative influence of social factors on naturally proceeding aging processes. Accordingly, the primary task, which is to reduce to a minimum the causes that lead to pathological forms of social aging coincides with more general social tasks relating to a reconstruction of society that would create normal human conditions for man's existence, including medical services. The right to health is also the basis for asserting one's right to life and especially to a long life, which realizes man's biological reserves more effectively and reduces the operation of factors producing early social aging to a minimum.

Here again the humanistic approach according to which the value of a long human life is self-evident and self-sufficient coincides with the social approach, which considers the increasing social significance of preserving highly developed forms of human individuality enriched by knowledge, life experience and wisdom. It would appear that increases in normal social longevity at the cost of limiting and eventually fully displacing pathological social aging is progressive both in relation to individuals and in relation to human society as a whole. It is true, of course, that achievement of that objective in full is possible only in the future.

Biological longevity is another matter. It refers to a life's *species time,* which is encoded genetically by evolution and pre-

---

[1] In particular, A. G. Vishnevsky traces various types of social forms of regulating longevity at various stages of human history as well as the prospects of reducing mortality at the threshold of the third millennium and beyond (See A. G. Vishnevsky, *Vosproizvodstvo naseleniya i obshestvo* /Reproduction of the Population and Society/, pp. 81-147). According to his data, in the Paleolithic period the average duration of life was 20 years. It was 30 years in the Middle Ages and is 70 years today.

supposes a succession of individual lives as the condition for its historical duration. This raises many new questions that are largely addressed to biology but they, too, cannot be considered independently of social and moral-humanistic issues that are defined by general solutions to the problem of the essence and meaning of human life, its duration, death and of man's immortality.

A scientific formulation of these problems will depend, above all, on advances in biology. In fact it is precisely studies in the essence of life and death in an overall biological context that created the prerequisites for studying these processes in man, including the biology of aging. And there is no need to turn to a distant past, for the method of formulating many questions that are relevant today dates primarily to the post-Darwinian period and is closely associated with the development of the evolutionary method in the science of life.

In that connection a substantial contribution was made by Ilya Mechnikov, a remarkable Russian scientist. A specific characteristic of his research is precisely a striving to establish relation between biological, social and humanistic approaches and a general philosophical formulation of the meaning of life. It is true that he has, nevertheless, emphasised the biological side, and this explains the naturalistic limitations of his overall position. This may be clearly seen in his *Studies in Human Nature* and especially his *Studies in Optimism,* in which the theme of the duration and meaning of human life, aging and death is developed in wide historical scope and yet often serves merely as a background for justifying his scientific-biological approach to the problem.[1] And while the latter has led to the establishment of gerontology as a science and has created the foundations of scientific research on problems of life and death as a whole, Mechnikov's social methods and moral-philosophical aspects receded into the background, and in a number of cases this made his views extremely one-sided. In his opinion, "the aim of human existence is to pass through a normal life cycle that

[1] I.I. Mechnikov, "Etiudy o prirode cheloveka. Opyt optimisticheskoi filosofii" (Studies in Human Nature. Essays on Optimistic Philosophy), in: *Akademicheskoye sobraniye sochinenii* (Academic Collection of Works), Vol. 11, Moscow, 1956; also *Etiudy optimizma* (Studies in Optimism), Moscow, 1964.

ends in a loss of one's life instinct and to a painless aging, reconciling the individual to death.[1] Mechnikov considered an age of approximately 100 or 120 and more years as a limit of normal human life—much more than the biblical age assertedly defined by King David (70-80 years). Mechnikov viewed premature aging as an illness that must be treated with the scientific methods of a future "rational macrobiotics". Above all, one should study "human nature", which Mechnikov viewed as "the basis of morality". Naturally, this led to both fair and sometimes unfounded, tendencious, above all, traditional-religious, idealistic criticism.[2]

In terms of modern scientific views Mechnikov's conceptions concerning the human organism's "harmony" and "disharmony" and his corresponding recommendations as to methods for the "treatment of old age" were greatly oversimplified. Nevertheless, *scientific* studies in human aging and death were given significant impulses precisely through a further development and critical reconsideration of Mechnikov's ideas.

This is also true of a number of Mechnikov's general views concerning man's mission and the purpose of human existence, the instinct of life and death, and his ideas for reshaping human nature and, hence, of morality in order to "attempt to reestablish the correct evolution of human life, i.e., to transform its disharmony into harmony (orthobiosis)."[3] According to Mechnikov, the correction of human nature's disharmonies with the

---

[1] I. I. Mechnikov, *Op. cit.*, Vol. 11, p. 29.

[2] It is characteristic that this was also the reaction of Lev Tolstoy, who did not spare biting remarks concerning Mechnikov's book *Studies in Human Nature:* they include both accusations of "scholarly stupidity" and conclusions to the effect that according to Mechnikov—"If one removes the rectum one will cease to think about the meaning of life...", as well as indignations at his asserted wish to "correct Nature", when instead, as Tolstoy believed, one should learn "respect for life". And on July 18, 1904, he noted in his diary that "Mechnikov invents how by removing one's rectum and scratching one's arse it is possible to neutralize old age and death", while "the greatest minds have sought to find an answer to the question not in one's arse but in man's spiritual essence (see Semyon Reznik, *Mechnikov,* Moscow, 1973, p. 157). Of course, this reflects not simply a failure to understand Mechnikov's ideas but their rejection in principle, since they contradicted Tolstoy's own moral-philosophical and religious views.

[3] I. I. Mechnikov, *op. cit.,* Vol. 11, p. 230.

help of scientific methods appears to be all the more possible since in the past aging was "more physiological and death was more natural". Beyond this the study of human nature makes it possible to establish the true aim of our existence and clarifies the meaning of genuine culture and genuine progress.[1]

Already in absolutizing the importance of "human nature" and the role of science in human life, Mechnikov leaves the firm grounds of reality. He was so carried away by utopian visions that he transformed his own life into a "laboratory", as it were, for studying his conceptions of man's aging and death. It seemed to him that as men advance in the direction of existence's "true objective", which is not life's immediate pleasures, but to "complete a normal cycle" they will largely have to renounce personal freedom for the sake of solidarity, control births, etc. Even though an aging of the population will take place so will its ability to realize social potentials through policy where elderly persons "who have acquired a greater experience and have preserved all their abilities because of a healthy physiological state will provide the greatest services to the future of the human race".[2] This will change the foundations of morality, reduce today's egoism, while the certainty that "science alone is able to oppose the sorrows that derive from human nature's disharmonies will inevitably lead to the development of education, and that alone will already increase human solidarity".[3] Man will subject his nature to active change making it more harmonious and closer to the ideal just as he has altered the nature of animals and of plants. These views culminate in a hymn to science and to its forthcoming role. "If we think of an ideal that is able to unify persons into a certain type of future religion then it can only be based on scientific findings. And if it is true, as is often said, that one cannot live without a faith then it can only be a faith in the universal meaning of science."[4]

Of course, knowledge may not necessarily become a "religion of the future", and the entire course of subsequent developments has shown the very great dangers that follow from any absolutizing of science at the expense of other factors in man's own

---

[1] *Ibid.*, p. 231.
[2] *Ibid.*, p. 238.
[3] *Ibid.*, p. 239.
[4] *Ibid.*, p. 239.

historical development and that of the human race as a whole. It is important to see the *decisive* role of the social, moral, ethical and humanistic factors that constitute man's essence, particularly in those questions that bear directly or indirectly on his life, aging and death. But at the same time such an emphasis has contributed to a concentration of scientific interest on studying these questions primarily within the framework of biology and subsequently this has exerted a positive influence and has led to the appearance of scientific studies whose volume and quality have far exceeded all that was produced in the course of many earlier centuries in the history of human knowledge concerning man's life, aging and death.

In that connection an important role was also played by a further development and critical rethinking of August Weissmann's ideas and, in particular, his teachings concerning aging and death as a result of growing differentiation of organisms in the course of evolution. Shmalgauzen emphasised the importance of age changes, especially as a result of the cessation of growth. It was his view that degeneration in "old age forms part of an individual's normal cycle, and death is the final link in a chain of life phenomena"[1]. As if continuing Engels' thought that was not yet widely known at that time concerning death as "an essential element of life",[2] he observed that "death is already inherently associated with life because any manifestation of life is only possible through the destruction of a specific quantity of living matter. Death represents the negative aspect of life, as it were. Without life there is no death and the latter's source is life itself"[3].

In that connection Shmalgauzen referred to Weissmann's teachings concerning immortality or, more precisely, the potential immortality of protozoans, in which, he stressed, "there is nothing resembling death. Death is a cessation of life and is always accompanied by the appearance of a corpse. Here we find neither. Everything continues to live and nothing perishes"[4]. In unfavourable conditions protozoans perish in large numbers,

---

[1] I. I. Shmalgauzen, *op. cit.*, p. 4.

[2] Frederick Engels, *Dialectics of Nature,* Progress Publishers, Moscow, 1974, p. 295.

[3] I. I. Shmalgauzen, *op. cit.,* p. 4.

[4] *Ibid.*, p. 5.

however, when divided into two, the specificity of the maternal individual is lost and is dissolved in daughter ones. Yet objectively, in Shmalgauzen's opinion, that loss of individuality cannot be compared to death. Death sets in as a result of the formation of a higher individuality—a rigorously defined form with a fully harmonious correlation of parts. It is preceded by aging, which is accompanied by a degeneration of organs and tissues, above all of its nervous system. In his view, the inevitability of a limit to growth also imposes certain constraints on life itself whose gradual approach is accompanied by phenomena of aging.

Like a number of other scientists, including Weissmann, Shmalgauzen associated longevity with an animal's general organization and, in particular, with the "cephalization factor", so that "the most intelligent animals perform the greatest vital work and live longer than others".[1] At the same time he did not believe that an increase in the mass of an animal's nervous system always leads to a longer life. That is possible only when a rigorous harmony is established in its entire organization. Moreover, in Shmalgauzen's opinion, it has not at all been proved that longevity is a biologically "useful" attribute. "We find it necessary to reject the presumption concerning the usefulness of a long life, and recognize that each animal species (and even each sex and each form) possesses its own characteristic longevity, which is not at all the one that is maximally possible, and is determined by the animal's entire organization. In certain cases we find lives that are strikingly short and that are clearly determined by a particular in-built disharmony in physiological constitution"[2].

It is in such a context that Shmalgauzen draws attention precisely to Weismann's view on longevity as an *adaptation* serving the reproduction of progeny, and on death as a "useful adaptation that prevents the reproduction of old, worn out individuals that could only reproduce a weak progeny"[3]. He concludes by observing that "biological longevity proves to be linked, above all, to the animal's reproductive ability and to biologically specific characteristics of the reproduction and the bearing of prog-

---

[1] *Ibid.*, p. 47.
[2] *Ibid.*, pp. 49-50.
[3] *Ibid.*, p. 50.

eny"[1]. In his opinion, "limitation of the life span through death is the result of restricting the growth and form of an animal by certain standards[2]. Limits on growth include limits on assimilation, and disruptions of metabolism and removing those phenomena or reducing their intensity will contribute to an enhanced longevity.

In concluding Shmalgauzen writes that Weissmann "viewed death as a secondary useful acquisition of animals in struggling for existence. Many other authors view death as an inevitable, final phase in the life cycle of all organisms, i.e., as a primary property of living matter. Up to a certain point both sides are correct"[3]. Shmalgauzen combines these approaches into a certain synthesis that unfortunately has been missed by a number of contemporary authors. In his opinion, "like its entire organization a life's duration is ultimately determined by the structure of sex cells, the composition of its 'genes'. In such a context, one of the conditions for maximal longevity is a rigorous harmony of the entire organization"[4].

It follows, according to Shmalgauzen, that the longevity of animals could be significantly increased, and if we do not see this today the obstacle lies not in the general properties of "living matter" but in the biological conditions governing an organism's existence. Referring to Weissmann, he associates this with a correct replacement of generations that the evolutionary process unquestionably requires.

As for man, whose relative longevity is longer than that of other mammals, Shmalgauzen considers that in his case a higher level of individuality is also accompanied by considerable resilience on the part of the organism. Accordingly, "a complete rejuvenation of the entire organism ... cannot be imagined without a loss of individuality, i.e., the very factor that constitutes our life's greatest value ... but in any event even without this, human longevity is already relatively long and can undoubt-

---

[1] I. I. Shmalgauzen, *op. cit.*, p. 51.

[2] *Ibid.*, p. 61. It is interesting to note that long before a similar view was expressed by Kant: "The very same mechanism that results in a man's life and growth eventually brings him death, when growth ceases." (I. Kant, *Sochineniya* (Works) in six volumes, Vol. 1, p. 98).

[3] I. I. Shmalgauzen, *op. cit.*, p. 85.

[4] *Ibid.*, p. 86.

edly be increased further. "But then," Shmalgauzen stressed, "we possess still another advantage over animals—the results of our creative activities do not perish with us but instead are accumulated to serve the well-being of future generations. So let our short lives be illuminated by our awareness that man's life is much higher than other lives and that only death has made possible the existence of the immortal creations of his spirit."[1]

These are the enlightened wishes with which this remarkable author and thinker concludes a work that is short, yet exceptionally profound and comprehensive and one that has exerted a considerable influence on modern conceptions of man's life, aging and death, and in particular, the development of modern gerontology.

Today there exist hundreds of different conceptions that emphasize and sometimes absolutize a variety of specific factors in the aging of the human organism. Many of them, moreover, simply modernize, on the basis of new biological (in particular, molecular-genetic) data, hypotheses that were proposed in the past. All these conceptions may be divided into two basic categories. According to the first, aging and death are programmed genetically, while according to the second they are conditioned by the emergence and accumulation of genetic damage that the organism does not have time to remove. But both conceptions and hypotheses usually link aging and death with evolutionary mechanisms, and this reflects the influence of the earlier approaches of such authors as Mechnikov, Weissmann and Shmalgauzen.

Such conceptions and hypotheses also display reductionist-biologizing tendencies that absolutize a variety of specific factors in the vital activities of human organisms. There is, however, a growing awareness of the need for an integrated, systems approach to the problem of human aging and death (according, for example, to the adaptive-regulatory theory of the Kiev school of gerontology) based on an understanding of their social essence which mediates biological factors and subordinates them to itself. But in that connection one occasionally finds a negative attitude towards the very idea that social conditions are a leading factor in gerontogenesis.

---

[1] *Ibid.*, p. 91.

As concerns the immediate future, emphasis is placed on the possibility and need for achieving, with the help of scientific methods, man's maximum species (biological) longevity which, in the opinion of a number of scientists, is in the order of 150 years. This is the aim of today's major efforts, even though one can already discern other aims and tasks that are more complex. One often even hears that we "are on the threshold of a new era when medicine will transform *Homo sapiens* into *Homo longevus,* when men and women will retain both their mental and physical liveliness in their mature years. And if this happens, we shall have to look upon life with totally different eyes."[1]

Which of these thoughts deserve particular attention? It seems to me that an interpretation of life must derive from real humanistic norms of man's awareness and behaviour, and a clear definition of the meaning of human life itself for the sake of which that life should be prolonged. While these norms do depend on social conditions they also exert a feedback relation on them. The dialectics of this process is given little emphasis in gerontological studies.

It is true that at present the desire to proceed from the dialectics of biological and social as well as individual and social phenomena is reflected in a number of theories of gerontogenesis, including efforts to rely on a functional subdivision of a person's life span into chronological, biological, psychological and social ages. It seems to me that it will not be very long when the very concept of "social", when applied to the duration of human life, will also be differentiated with due regard for personal (including moral-humanistic) awareness and behaviour norms. For no matter what macrobiotics (the science of increasing longevity) can promise us whether today or in the future, Seneca's wise words to the effect that the best way to increase the duration of life is not to make it shorter, will continue to hold true.

And there is also Kant's following observation, "he lives long-

---

[1] J. Kurtzman, Ph. Gordon, *No More Dying,* Los Angeles, 1976. The authors consider various artificial methods for prolonging life including transplantations, bionic technologies, cryobiology and gene technology. Unfortunately they in effect exclude the leading role of sociomoral factors in gerontogenesis.

est who takes the least trouble to prolong life but carefully avoids shortening it through interference with our beneficent nature."[1] It should be noted that Kant did not simply limit himself to everyday observations. He produced a special treatise entitled "On the Powers of the Spirit to Be Master of Painful Sensations through Sheer Resolution". This was a reply to a work that was extremely popular in those days by C. W. Hufeland, a well-known German medical doctor, *The Art of Prolonging Human Life.* Kant's treatise contains many statements that today will cause readers knowledgeable in medicine to smile, but it contains one feature that is important and instructive; he ascribes a special significance to spiritual and, in particular, to moral factors in prolonging human life. It is true that in the long run for the philosopher the "moral-practical philosophy is at the same time a universal medicine,"[2] which even if it does not cure "neither can it be absent from any prescription."[3] Moreover, while agreeing with Hufeland in defining dietetics as the ability to avoid illnesses and thus prolong life (but not to make it pleasureful) Kant refers to stoicism as a principle of dietetics and, hence, an element of practical philosophy "not only as the science of virtue but also as the science of *doctoring.* The latter becomes *philosophical* when the mode of life is determined only by man's power of reason to be master of his sensual feelings through a rule of conduct adopted by man himself."[4] This also extends to philosophizing which operates as a means for displacing unpleasant sensations and producing an exitation of spiritual life in those persons as well who are not essentially philosophers. This interest, in Kant's view, "does not let the vital force stagnate. On the contrary, philosophy, whose interest is for the whole of the final goal of reason (which is absolute unity), has a feeling of power that can to some extent compensate for the bodily weakness of age through a sensible evaluation of life's worth."[5] Kant also notes, however, that any direct interest in science, too, operates in a similar manner, and that a person who immerses himself in science, "is to that extent

[1] Immanuel Kant, *Briefwechsel,* Vol. 1, Berlin-Leipzig, 1922, p. 344.
[2] Immanuel Kant, *Werke,* Vol. VII, Berlin, 1916, p. 412.
[3] *Ibid.*
[4] Quoted from *Social Sciences,* No. 2, 1984.
[5] *Ibid.*

a philosopher and enjoys the beneficent effect of such an excitement of his powers in a rejuvenated life extended without exhaustion."[1] Moreover, Kant slyly adds, that a preoccupation with trifles produces nearly the same results: a carefree existence leads such persons to achieve a ripe old age fearing only that they may die from boredom.

C. N. Parkinson and H. Le Compte have sought to combine a number of interesting thoughts into a "law of longer life."[2] In spite of the fact that they continue to adhere to one-sided biologizing and Malthusian views, they state the following rational idea. Where there exist many ways for prolonging life, one should remember that the very question whether to live or to die greatly depends on the state of one's mind. We die at least in part because we have lived sufficiently long: we live because here is *still something that we must do*.[3] I particularly stress this last thought, for it clearly links the duration of human life to personal guidelines including an understanding of life's meaning and objective.

In such a context Engels' views expressed in a letter to Friedrich Sorge on the day following Marx's death are indicative: "Medical skill might have been able to assure him a few more years of vegetative existence, the life of a helpless being, dying —to the triumph of the physicians' art—not suddenly, but inch by inch. Our Marx however would never have borne that. To live, with all the unfinished works before him, tantalized by the desire to complete them and unable to do so, would have been a thousand times more bitter to him than the gentle death that overtook him."[4]

Great importance also attaches to the development of concepts in which the leading role of social factors would be established

---

[1] Immanuel Kant, *Werke,* Vol. VII, p. 412.

[2] C. N. Parkinson, H. Le Compte, *The Law of Longer Life,* Troy, Alabama, 1980.

[3] This was already noted by Cicero who believed that "age does not possess defined limits and one continues to live correctly in old age so long as one is able to both fulfill effectively the burdens of one's duty and have contempt for death. This is why old age has a greater spiritual strength than does adolescence." (Tsitseron, *Izbranniye sochineniya* (Cicero, Selected Works), Moscow, 1975, p. 380).

[4] "Engels to Friedrich Adolph Sorge in Hoboken, March 15, 1883", Marx, Engels, *Selected Correspondence,* pp. 339-340.

in a socio-biological context corresponding to man's actual nature as biosocial being. In particular, a scientific understanding of specific features of the moral-ethical foundations and meaning of human life, aging, and death should be of enormous methodological and axiological importance. Many facts and tendencies point to this.

The future science of man as a humanized and truly *humane* science—will not only study biological approaches to the prolongation of human life but will also define its meaning in ways that combine biological components dialectically with social and ethical factors and the humanistic understanding of life and death. That measure is closely linked to an optimal realization of man's essential forces, and accordingly, it is not the duration of individual life itself that will serve as the objective of science and society and especially individuals but rather an unfolding of human nature's own wealth. The degree of a person's involvement in the human race's collective life and his participation in realizing the idea of an unlimited development of man as a social being will determine individual parameters of human life that accord with biological possibilites.

This will also require a qualitative change in general conceptions concerning the meaning of human life and one's attitude towards death. Today, when science is asked to prolong human life, the objectives of life itself are usually defined in very diverse ways, and its time limits, too, are viewed differently, and not always made dependent on social, ethical, and humanistic factors. This is why, in particular, one even hears statements to the effect that it may eventually be possible to prolong human life to 1000 years or even more, and possibly ad infinitum.[1]

But what does this look like on a moral-humanistic plane? Will the human race always strive to a maximal prolongation of *individual* life and, moreover, to immortality? Or will it find other solutions and will its consciousness change the very understanding of the meaning of human life to such an extent that the individual will not view himself as being isolated from the human race whose needs and interests will be perceived by him

[1] I am referring not to some kind of futurological euphoria, but also to forecasts made by a number of scientists. In particular, Arthur C. Clarke concludes that man may become immortal already by the year 2090 (Arthur C. Clarke, *Profiles of the Future,* New York, 1962).

as dominant? In such an event the life span of an individual person will be determined by concrete historical conditions and by the possibilities and needs of society that coincide with personal needs and possibilities. And, consequently, the consciousness determined by the being of this future society will exert an active influence on both society itself and each individual person.

It is difficult to know how science will in fact solve future problems in prolonging human life. But one thing is clear: the path to such a solution presupposes a reliance on a diversity of methods that will reflect complex interactions within the human organism as an integral system. It is even more difficult to say today what concrete time parameters will define the individual biological life of persons, but then it is hardly important for us to do this on behalf of those who will follow us.

Today many studies have shown that the level of intellectual functioning achieved at maturity may be preserved until very old age.[1] This is why the basic task appears to be to apply that priceless gift as fully as possible without initially specifying vague objectives for the human race and for an individual person.[2] One cannot know how the very possibility of reaching beyond life's former boundaries would influence individuals—in a social, psychological and moral-ethical sense. For that presupposes substantial changes in the human organism with the help of "homotechnology", and that in turn, may produce a loss of human individuality and of personal identity.[3] But a threat is also

---

[1] I. M. Hulicka, *Empirical Studies in the Psychology and Sociobiology of Aging,* New York, 1977; R. M. Henig, *The Myth of Senility,* Doubleday, New York, 1981.

[2] In terms of philosophy that problem has been partly reflected in polemics between Bernard Shaw, who in his drama *Back to Methuselah* declared that in order to solve many of the human race's social problems persons must live at least 300 years, and Karel Čapek, who has shown in his play *The Makropulos Secret* that such a life span is not only not ideal but even clearly undesirable. The heroine in Čapek's play chose death when she realized how senseless and empty her long and aimless life was.

[3] In the opinion of some scientists, this technology includes the transplantation of organs and also their artificial replacement with the help of new bionic technologies, psychosurgery, and even the creation of "biocyborgs". Consideration is also given to the idea of avoiding death through cloning individuals. Unfortunately, realities are often suppress-

presented not only by gerontophobia, but also by gerontophilia, as by any interruption in the historical movement of generations, a conservation of what is already achieved and a frightening prospect of its extrapolation over many centuries. It is unlikely that any person who is truly worthy of eternity would agree to become some kind of eternal embodiment and standard of "man generally", and in this way impose himself on the future, as it were, thus destroying the whole attractiveness of novelty and mystery in the "absolute movement of becoming".

Actually this was already well shown by Jonathan Swift by citing the example of the "elite" inhabitants of Laputa—the unhappy Struldbruggs, doomed to endless life and envying the death of other old men. And Goethe's Faust rejects suicide not from an egotistic desire to prolong life, but from a love of others, in order to participate in the human race's common fate while preserving youth. Today this idea is again being resurrected in so-called *juvenology* whose guidelines, I believe, must still be subjected to serious tests not only from a biological but also from a moral and humanistic point of view. In any event, I find that I.V. Davydovsky's *gerontological* ideas are much more firmly grounded and more attractive. He believes that "longevity and the problem of an *active creative old age* associated with it—is something that is more real than a boring immortality".[1] I find that such a way of stating the problem fully accords not only with current scientific prospects and those of the immediate future, but also—and this is the most important factor—with moral-humanistic principles operating as major, and later perhaps even as decisive, regulators of human longevity. B. Ts. Urlanis, a Soviet demographer, has observed that "we value life and treasure it precisely because it is not infinite. The most important thing is that its thread not be broken until the entire cloth has been woven, that the light not be extinguished so long as there is fire in it."[2]

The humanistic principle according to which one should "live

---

ed here by "scientific mythology" exploiting precisely the *possibilities* of science that are not related to the meaning of human life.

[1] I. V. Davydovsky, *Gerontologia* (Gerontology), p. 19.

[2] B. Ts. Urlanis, *Evolutsia prodolzhitelnosti zhizni* (The Evolution of Longevity), Statistika, Moscow, 1978, p. 302.

not only for oneself but for those who will follow us" calls for a scientific and philosophical interpretation of the meaning of human life, and of the human race's general social and moral progress both today and in the future. The human race has much to study and to overcome along that road.

In recent years such reflections have often turned to the past and sometimes consider restoring the religious-idealistic teachings now forgotten or half-forgotten by history. This does not only concern traditional religious dogmas relating to the immortality of the human "soul" and to conceptions of "resurrection" and "life beyond the grave" (as it is said in the Bible, the body will return to the earth from which it has come, while the spirit will return to God who has given it). And not only theologians, but also some geneticists, psychophysiologists, and even cosmologists seek to somehow "integrate" old dogmas with the latest findings of science. Today the reflections of a number of contemporary scientists concerning the future are turning to conceptions and ideas that not only do not correspond to progressive historical-philosophical traditions but also do not accord with Nature and social reality and with the outcome of the development of scientific knowledge. And even though there is an increasing general recognition today in scientific and philosophical approaches to the questions of human life and death, of the decisive role of socio-ethical and humanistic factors in human development, this does not assume the absurd forms that they have been given both in orthodox religious idealistic conceptions and in corresponding conceptions that have been modernized in accordance with new scientific findings.[1]

Let us now turn to the history of thought and examine how the most complex questions of human existence were defined and solved, specifically in Russian science and culture.

[1] In recent years theologians of the most various orientations have been intensively engaged in such activities. In particular, in an address in Turin on April 13, 1980, Pope John-Paul II, referred to rational psychological foundations of the fear of death. In his opinion, the wish to live is met and satisfied in accordance with the Christian teaching of "resurrection".

2. Illusions of "Eternal Life" and "Universal Resurrection". The Moral-Philosophical Meaning of Death and of Human Immortality (Examples from the History of Russian and World Thought)

In the history of Russian philosophy these ideas have found a somewhat original embodiment in Nikolai Fedorov's "project for a common cause".[1] The work of that Russian thinker was consonant with the spiritual searchings of such writers as Lev Tolstoy, Fyodor Dostoyevsky and Vladimir Soloviev who dealt with the problems of the meaning of human life, death and immortality. What are the specifics of Fedorov's "project" for resolving these problems? Why do a number of authors today consider that this "project" should be interpreted in terms of modern science and practice?

In *The Philosophy of the Common Cause,* Fedorov raises the issue of "brotherhood or kinship, of the causes of the non-brotherly, non-kindred, i.e., non-peaceful, state of the world and of the means for restoring kinship."[2] He addresses himself on behalf of "non-scientists" to scientists, both spiritual and secular, believers and non-believers in order to organize "joint efforts" to help regulate "all worlds through all resurrected generations."[3] He considers that "as soon as pride in the heroic deeds of forefathers will be replaced by a feeling of grief for their death, as soon as we will view the world as a cemetery and Nature as a mortal force, the political issue will be replaced by a physical issue and one, moreover, that will not be distinct from the dimensions of astronomy, i.e., the Earth will be recognized as a celestial body, and the stars as earths."[4]

The view that there is a "multiplicity of worlds" reflects an ancient cosmological thought and its creation should not be ascribed to Fedorov. By the way, the conclusion that Fedorov

---

[1] *Filosofiya obshchego dela. Statyi, mysli i pisma Nikolaya Fedorovicha Fedorova* (The Philosophy of the Common Cause. Articles, Reflections and Letters of Nikolai Fedorovich Fedorov), Vol. I, Verny, 1906; Vol. II, Moscow, 1913; also N. F. Fedorov, *Sochineniya* (Works), Moscow, 1982.

[2] *Ibid.*, Vol. I, p. 1.

[3] *Ibid.*, p. 9.

[4] *Ibid.*

makes is important to us. In his opinion, "a humanist who views himself as a *man* and is proud of that name has apparently not yet reached the measure of Christ's age, has not yet become a *son of man*; and all those who have rejected in our own time the cult of fathers have lost as a result the right to call themselves sons of men. . ."[1] Yet, according to Fedorov, the latter's existence as "resurrectors" is not only of exceptional importance, but is absolutely necessary if life's objective is to turn Nature's blind force into the one that is regulated by the minds of all resurrected generations. . ."[2]

We see that Fedorov defines the meaning of life in an extremely original way: it is opposed to a humanism that affirms man as a personality rather than simply as part of a larger whole—as a "son of man". What follows from such a definition? That all means, including science, must be subordinated to an "ultimate objective", namely, that of uniting "all in the common cause of turning Nature's blind force into an instrument of the human mind in order to return what has been displaced."[3] This is the basis of Fedorov's criticism of materialism and positivism, and more generally, of the science of his own times, which did not see that objective.

Fedorov's strange "ultimate objective" is not simply utopian, but also senseless, and a far cry from genuine humanism. Accordingly, no references to science (including modern genetics, as some modern authors do) will bring this idea closer to reality.

Yet with Fedorov himself, that absurd idea seems "fully logical", for he bases his position on certain non-traditional interpretations of Christian dogmas which science, in his opinion, should also follow. In his view, "a rejection of religion consists not in the true or good but in the evil application of blind force, in subordinating oneself to it under the pretence of ruling over it, in submission to it in sex selection (in manufacturing) and in natural selection (various forms of destruction). Serving the God of our fathers consists in turning blind, mortal force, through regulation, into life-giving force. Unlike the exploitation

---

[1] *Filosofiya obshchego dela. Op. cit.*, p. 11.

[2] *Ibid.*

[3] *Ibid.*

and utilization of nature, i. e., in contrast to its plunder by prodigal sons for the sake of their wives, which results in exhaustion and death, regulation leads to a restoration of life. Not only idolatry constitutes a distortion of religion, but worship of thought or idolatry, too, is such a distortion; while philosophy as the work of scientists who have isolated themselves from others is the greatest distortion of religion."[1]

In that passage we see that even the reasonable thought concerning "regulating Nature" is given a religious form, and aimed, in particular, against science and against philosophy, which for various reasons the author of *The Philosophy of the Common Cause* treated with particular disdain. Moreover, he viewed as the major flaw of *all* philosophical systems the absence of the very idea of resurrection or non-recognition of the "common cause". And while one cannot deny that Fedorov has contributed a number of critical characterizations, for example, of Nitzsche's teaching concerning "supermen", nevertheless, his evaluations of many aspects of the moral philosophy of Kant, Vladimir Soloviev, and especially of Lev Tolstoy are quite false and one-sided. And this is only natural, since in the history of philosophy, and, in particular, that of the materialist systems that Fedorov characterized with so much ill-will, a conception of life and death was developing that opposed religious and idealist conceptions, and Christian myths concerning the "immortality of the soul" and "resurrection" were criticized.

There is no need in the present context to examine closely his idea of a "universal resurrection". But it should be described, even though briefly, in order to make clear the artificiality of attempts to link it with modern science, with concepts of heredity and with current biomedical research concerned with increasing the species longevity of human life.

According to Fedorov, it is precisely "God the Trinity" that bears within Himself a "need for resurrection". Moreover, "Nature, being aware of losses in us, the living generation, restores us through ourselves, through our knowledge and activities, restores these losses, i.e., all dead generations, from itself. It is only in its full complement, in the totality of all generations

---

[1] *Ibid.*, p. 48.

that the human species can enter into the unity that was promised to it, into communion with the trinity, enter into it, as if into one's own frame."[1] Furthermore, since Fedorov views the organism as a machine, he refers to a "reproduction" of the human race from "simplest elements" ("assemble the machine and awareness will return to it") "passing through all the individualities through which these elements have passed", "in accordance with the laws of heredity", using "psycho-physiological diaries", etc.[2]

A number of contemporary studies of Fedorov interpret these ideas as "scientific prevision", although with a reservation to the effect that they also express his vulgar-materialist views, combined (assertedly only formally) with religious conceptions relating to the "Trinity". In such a context emphasis is placed on differences between traditional Christian ideas and Fedorov's conceptions concerning "universal resurrection". Yet that difference (for Christianity, too, believes not only in an immortal soul, but in a future "universal resurrection") is incomparably smaller than is the general distance of these conceptions from science and materialism. And it is not surprising that Fedorov criticizes science, and, in particular, the views of Mechnikov, even more sharply than does Tolstoy. It is true Fedorov refers to "Western science" which developed "within a commercial and industrial organism", and not to the science of the future that he postulates, striving to realize the "common cause", developing within an "agrarian environment", and having already become a science not of "decomposition and killing but a science of combining and restoring."[3] The former, in Fedorov's view, essentially representing a paganism, but only secularized, that began with the age of the "so-called Renaissance", is characterized by a superstitious awe for natural facts including death. And yet, in his view, "death is simply the result or expression of minority, or a non-independent, non-original life of an incapacity for mutually restoring and supporting life. People are still not fully grown and only half-beings; but a fullness of personal being, personal perfection is only possible when there is universal perfection. Majority is painlessness and immortality;

[1] *Filosofiya obshchego dela. Op. cit.,* p. 85.

[2] *Ibid.,* pp. 288, 303, 320.

[3] *Ibid,* p. 331.

but without a resurrection of the dead the immortality of the living is impossible."[1]

In short, if you wish to be immortal you must become a "re-surrector" and that, Fedorov believes, is "a total triumph of law over physical necessity".[2] This is also the outcome, in his view, of the need to overcome hunger and to resolve what he describes as the problem of "sanitation". "By turning the unconscious process of birth and also of nutrition into action, namely, into a universal resurrection, the human race turns all worlds, through recreated generations, into means of existence. Only in this way is it possible to resolve Malthus' formula of the opposition between reproduction and means of existence. On the other hand, it is only in this way that the human race will also get rid of universal mortality, which has originated as an accident from ignorance, and, hence, from powerlessness, and which through heredity has become an inborn epidemic disease by comparison with which all other epidemics may be viewed as sporadic. Mortality has become a universal organic vice, flaw, a deformity that we have ceased to recognize and do not consider as either a flaw or a deformity."[3]

Fedorov reproaches philosophers with the fact that some of them do not even wish to recognize death as an evil. "Death is a property, a state attributable to some reasons but not a quality without which man ceases to be that which *he is and that which he must be*."[4] That is the specific way Fedorov formulates the problem of the essence of death. Is it really creative, as some modern authors seek to persuade us? Of course not. We are convinced of this by turning not only to the history of philosophy and of scientific thought but also to certain modern conceptions that have developed independently of those of Fedorov.

Nor can still another question be avoided. Fedorov believed that "for its realization the natural cause, i.e., resurrection, requires two unifications—unification of the external, which may take place through autocracy, and of the internal, through orthodoxy; and this will be a unification of all sensible beings for the cause of gaining a knowledge of the insensible force

[1] *Ibid.*, p. 91.
[2] *Ibid.*, p. 131.
[3] *Ibid.*, p. 277.
[4] *Ibid.*, p. 287.

which, while giving birth, kills, and of regulating the insensible force by sensible being (sons)".[1] Of course, these conclusions of Fedorov may be interpreted differently or simply ignored, as is his proposal to replace the question of poverty and wealth by the question of death and life.[2] But this will hardly lead us to objective assessments of those of his ideas which are being publicized so indiscreetly today. Fedorov associates the idea of immortality with a deep restructuring of human nature, even though he believes that this will not be accompanied by changes in man's essence: man will simply be more himself than he is at present.[3] As a result, "the ultimate goal of the life of sensible beings is to become initial cause of themselves and to become like the initial cause, the divine original cause".[4]

Similar ideas are also being discussed today. But, of course, their scientific examination calls for different reference points than those that *The Philosophy of the Common Cause* can provide: "Life is a good, death is an evil. The return by the living of life to all those who have died for the sake of immortal life is a good without evil. The recreation from the earth of all the dead, their liberation from the power of the earth and the subordination of all lands and all worlds to the resurrected generations—that is the loftiest cause of the human race, its highest duty and at the same time its supreme good."[5] Such an interpretation of the highest cause, duty and good is hardly likely to be able to inspire the human race, for it is based on absurd premises that do not cease to be absurd because that understanding contains individual rational ideas and insights.

---

[1] *Filosofiya obshchego dela. Op. cit.*, p. 422.

[2] *Ibid.*, p. 422.

[3] "His organs will also include those modes of aeronautics and etheronautics that will make it possible for him to travel in outer space and obtain materials for constructing his organism. Man will then bear within himself the entire history of discoveries, the entire course of that progress; he will contain both physics and chemistry, in short, all of cosmogony, however, not in the form of mental images but in the form of a space apparatus that will make it possible for him to truly become a cosmopolite, i.e., to be everywhere in succession, and then man will become a truly enlightened being." (*Filosofia obshchego dela.* Vol. I, p. 318.)

[4] *Ibid.*, Vol. II, p. 48.

[5] *Ibid.*, p. 122.

Absurdity is a hypertrophy and shift in the realization of phenomena that actually exist.

Fedorov's rational ideas concern a conscious regulation, on a cosmic scale, of all forces of Nature—material and spiritual—"by beings that are sensible and moral, and jointly working for the common cause".[1] But they become nonsensical when the concept of a "common cause" is itself defined as a "universal resurrection", within which man's life and death are interpreted as mutually exclusive opposites, as sharply opposed forms of good and evil.

Fedorov constantly stresses that mortality "is not an unconditional necessity", and may be regulated by the mind.[2] From these positions he criticizes both "the worn arguments of materialism",[3] and all "reformers, including socialists", who seek to "make mortals happy". He presented his idea of a "psychocracy" as the "opposite to Marxist materiocracy".[4] In moral-philosophical terms Fedorov defined that position as "neither egoism nor altruism, but kinship". This meant to live "neither for oneself, nor for others", but for a "universal resurrection" in which immortality, too, becomes a legacy not of individual persons but of the entire human race.[5]

In spite of a certain positive meaning, on the whole such an approach imposes conceptions that enter into deep contradictions with the humanistic tradition of science and philosophy.[6] This can also be seen from the fact that Fedorov violently attacks Mechnikov's teachings concerning orthobiosis, and especially Lev Tolstoy's moral-philosophical ideas. He reproaches Tolstoy with the fact that in asserting that "death is not a bad thing" "he does not believe in miracles and does not recognize logic". In doing this, Fedorov rejects those moral grounds that underlie Tolstoy's assertion.[7]

---

[1] *Ibid.*, p. 180.

[2] *Ibid.*, p. 203.

[3] *Ibid.*, p. 209.

[4] *Ibid.*, p. 271.

[5] *Ibid.*, p. 201.

[6] It is not accidental that Fedorov defines humanism as "mental and moral groping in the dark". (*Filosofiya obshchego dela, op. cit.*, Vol. II, 199.)

[7] N. F. Fedorov, *Sochineniya* (Works), pp. 637, 640-641.

Philosophy's humanistic tradition was always directed against attempts of theologians to ascribe to religion alone the right to solve questions relating to the meaning of human life, death and immortality, by turning to God, who, it is asserted, "overcomes" the fear of death by ensuring personal immortality in the future. It is clearly reflected in Lev Tolstoy's moral-philosophical reflections on the problems of death and human immortality that concerned him so much. In many ways his ideas are opposite to those of Fedorov, and while they, too, rest on a religious-idealistic basis, they were given, especially in his literary works, a deep life-like content that stood in opposition to the Orthodox Christian teachings of death and "resurrection" and also to Fedorov's corresponding concepts. This is why they continue to have an enormous moral influence on man today, and on his reflections concerning death and immortality, and why representatives of different philosophical systems are turning to them.

What is their attraction for modern man, whose consciousness is being enriched by concepts and images stimulated by the scientific and technical revolution, gerontology and juvenology, projects of a radical restructuring of human nature and an infinite increase in man's longevity? The very shortest answer would be—*their extreme humaneness,* that is, an organic union of thought with the feelings of a unique personality as connected with other persons and the human race as a whole. Tolstoy shares with others the torment of his reflections on the meaning of life and of death, and his readers realize that there is no final answer, nor can there be one, for its meaning lies in life itself, in its eternal movement as man's own becoming.

In contrast to Fedorov's views on the problem of man's death and immortality, which belittle the meaning and task of human life, Tolstoy, who also did not reject the "teachings about Christ", boldly affirmed a humanistic vision of living man's self-value. He gave him not merely a "consolation" by removing the fear of death, but also a wise genuine awareness of the meaning of the life of man as a social and moral personality. In the works of both Tolstoy and Dostoyevsky, and even in those of Soloviev, the image of Christ appears as a moral ideal rather than as the symbol of a cult. In Tolstoy's teachings it plays a prescriptive-regulating role rather than an ontological one. According to Tolstoy, "true religion is a relation to surrounding infinite life, established in harmony with man's mind and knowledge that links his life to that infinity and guides his actions"[1]. This is why many of Tolstoy's thoughts also strike a deep resonance within us, and generate our own reflections and experiences.

The hidden meaning of what takes place within man as he passes from life to death has been shown by Tolstoy with the full force of his powerful talent. He presented this in an extremely sharp and clear, yet paradoxical form in an article entitled "Concerning Life": "The only reason why man dies is that the well-being of his true life can no longer increase in this world rather than because he has a lung disease, or cancer, or he has been shot at, or because a bomb has been thrown at him".[2]

We find a preoccupation with the inevitability of man's biological death throughout all of Tolstoy's literary works. But it is also combined with an affirmation of man's spiritual immortality. This is embodied in life itself, in what constitutes its basis, "a specific attitude on the part of each being towards the world"[3]. Death is frightening to he who "does not see the senselessness and destructiveness of his personal, lonely life, and who thinks that he will not die... I will die like everyone ... but my life and death will have meaning both for myself and for everyone"[4].

Thus, Tolstoy also extends life's moral idea to death. This is why he considers that "a person may have died but his attitude to life continues to influence others, and even much more than when he was alive, and that influence continues to increase as intelligence and love also grow, and like everything that is alive, never ceases and knows no interruptions".[5]

Tolstoy has clearly illustrated such an approach to the problem of man's life, death and immortality in his article "On Suicide" (1900) and also in "The Green Stick" (1905). "So long as a person is alive he can become more perfect and serve the world, but he can serve the world only by becoming more perfect, and can become more perfect only by serving the

---

[1] L. N. Tolstoy, *Polnoye sobraniye sochinenii* (Collected Works), Vol. 35, p. 163.

[2] *Ibid.*, Vol. 26, p. 422.

[3] *Ibid.*, p. 401.

[4] *Ibid.*, Vol. 23, p. 402.

[5] *Ibid.*, Vol. 26, p. 413.

world".[1] If a person "lives, forsaking his well-being for that of others, already in this life he enters into that new relation to the world for which there is no death and whose establishment is for everyone this life's cause".[2]

Tolstoy's philosophical conclusions concerning man's death and immortality are also embodied in the literary meaning that he lends to these problems. This is shown in a remarkable way in *War and Peace* and especially in *The Death of Ivan Ilyich*.[3] Let us recall the description of Andrei Bolkonsky's transition from life to death after being severely wounded in the Battle of Austerlitz. "There was no longer anything above him except the sky which was high, not clear but still immensely high and infinite, that he had not seen before and he was happy to know it and to find calm in it..." And everything seems insignificant by comparison with "this high, just and kind sky": grandeur, life itself, "whose meaning no one can understand", and even more insignificant—death, "whose meaning none of the living could either understand or explain".

The constant reflections of Tolstoy's heroes "concerning life and concerning man's mission" ("so that my life would not proceed for me alone but would influence everyone, and that all of them would go with me") make death itself for them all a phenomenon that is basically spiritual and moral, and also associated with extremely complex processes that are only inherent in man as a personality. This is especially deeply reflected in the description of Andrei Bolkonsky's state following his second and extremely serious wounding in the Battle of Borodino. As always, Tolstoy shifts aside ironically and disdainfully everything concerned with the medical aspect of the matter and concentrates

---

[1] L. N. Tolstoy, *op. cit.*, Vol. 18, p. 220.

[2] *Ibid.*, Vol. 26, p. 415.

[3] In a certain sense that story also stimulated Heidegger's and Sartre's reflections on these problems, although it would appear that this refers primarily to that part that specialists consider to be a highly accurate psychological description of the process of a person's death. For in their own philosophies we find many features against which Tolstoy had warned: "But avoid what the entire European world is now doing: living without considering their life intelligent, engage in activities that they do not view as intelligent, have no faith in their mind, and live in disharmony with it." (L. N. Tolstoy, *op. cit.*, Vol. 23, p. 450.) We shall return to this later.

on the spiritual-moral symptoms of Andrei Bolkonsky's death: "Suddenly it happened", "an alienation [appeared] from everything worldly that was frightening to anyone alive. Apparently, he now had difficulty in understanding everything that was alive; but at the same time he sensed that the reason why he did not understand living things was not that he had lost his power to understand but because he understood something else, something that the living did not and could not understand and that absorbed him completely." Prince Andrei calmly awaited something that was "terrible, eternal, unknown and distant" and that was death. For Tolstoy death is associated with discovering the "principle of eternal love" of all and everything and with the desire to sacrifice oneself for its sake. But as he was increasingly imbued with this principle, Andrei Bolkonsky "was renouncing life, and in so doing he was destroying even more effectively that frightening barrier that in the absence of love stands between life and death". We see the "last moral fight between life and death, in which death prevailed... It came in, and it was *death*, and Prince Andrei died". But according to Tolstoy, it is also an "awakening from life", and "that is how it must be" and "that is good".

In later years Tolstoy deepens his spiritual and moral interpretation of human death and associates it with highly personalized processes and emotional experiences. In *The Death of Ivan Ilyich* we follow step by step the death of a forty-five-year old person whose short life history was "simple, ordinary, and appalling", even though it had proceeded not only "easily, enjoyably and decently but even elegantly". But suddenly, he fell ill and "something frightening and new took place within him that was so significant, so much more significant than anything Ivan Ilyich had experienced during his life". Time passed and "in examining anatomically and physiologically the details of what the doctor believed was taking place within him, he understood everything".

He understood that "it isn't a matter of caecum and kidneys, it is a matter of life... and death". Ivan Ilyich saw that he was dying, but he could not understand this and was in constant despair. That despair was not helped either by the syllogism that "Caius is a man, men are mortal, and therefore Caius is mortal", nor by other thoughts in which he was hoping to find support. Not only physical pain was a torture for Ivan Ilyich but also the lie to which he was being subjected before he was to die. But above all, he understood that he had not lived as he should have, and these moral sufferings were more terrible than the physical ones; this was what tormented him most.

With a masterly care for details Tolstoy describes Ivan Ilyich's dying as a person's "moral sufferings" that ended only when he "saw a light and saw that his life was not what it should have been, but that this can still be corrected". And even simple human pity for those who are close to him brings him relief: "And presently it became clear to him that all he had been tortured by and unable to throw off, was now falling away of itself, falling away on two sides, ten sides, all sides at once." He felt sorry for them, he must do something to ease their pain. He must relieve them and himself of this suffering. "'How good and how simple' he thought..." That awareness prevailed over pain and over the habitual fear of death, and even over death itself, which was being replaced by light, joy, and . . . true death sets in.

The moral-humanistic significance of such a perception of death, that is closely associated with arriving at the meaning of a person's life and of his historical immortality is extremely important. This is why its influence on the development of moral-philosophical thought concerning that "eternal" theme has been so considerable.

Although he does this in a somewhat different form, Dostoyevsky, too, develops these ideas, especially in his *Crime and Punishment* and *The Karamazov Brothers* and also in some of his journalistic works. In spite of many specific characteristics in his interpretation of man's death and immortality by inheriting and further developing the humanistic historico-philosophical tradition, these thinkers stimulate today's philosophical reflections and spiritual and spiritual-moral experiences connected with this problem.

That tradition originates in ancient times. In ancient Egyptian beliefs, in Buddhism, Taoism and also in ancient Greek idealist philosophical systems (particularly those of Socrates and of Plato), attempts to morally overcome the fear of death were associated with conceptions of a "life beyond the grave" and of "the immortality of the soul". These ideas were further developed by Aristotle in his treatise *On the Soul.* Many conceptions, including materialist ones, derived from a view of existence as an eternal cosmic circular motion and viewed man as a part of that process. This is why according to Epicurus, for example, death has no meaning for us, since whatever decomposes into atoms loses its ability to feel and, hence, has no relation to us. Lucretius explained this by postulating that man's mind dies together with him and cannot know what happens "after death"; death, in his opinion, is an inevitable event as old persons yield their place to young people. We did not think of the possibility of eternal life until the moment when we appeared on this earth, so why should we think of it after death? The suffering that we think will begin after death is in fact the suffering that only exists in the present. It was Lucretius' view that together with other considerations this can reconcile man with death.

Similarly many of Cicero's maxims affirm a moral force in facing death. In particular, they are reflected in his *Tusculanae Disputationes* whose first part entitled "On Feeling Contempt for Death" seeks to justify the proposition that death is an evil and an "eternal misfortune".[1] In the dialogue entitled "Cato the Elder or On Old Age", Cicero writes that "more generally it seems to me that whoever has been satiated with everything has also been satiated with life itself ... and his time of death has arrived".[2] A belief in the soul's immortality[3] leads Cicero to the conviction that the dead "are alive, and live a life, moreover, that alone deserve the name of life". But he also believes that: "If I am mistaken in believing in the human soul's immortality, I commit a fallacy and do not wish to be deprived of a fallacy that gives me pleasure while I live; if, on the other hand, I will not feel anything after death as some lowly philosophers believe, then there is no reason why I should fear the ridicule of dead philosophers. If we are not destined to become immortal then it is still better to die at the proper time: for nature establishes a measure for life as for everything else, and old age is the last scene in the life's drama;

[1] Marcus Tullius Cicero, *Izbranniye Sochineniya* (Selected Works), Moscow, 1975, p. 210. In this work Cicero examines many conceptions of man's death and immortality that are expressed in the philosophical systems of antiquity.

[2] *Ibid.*, p. 382.

[3] *Ibid.*, p. 382.

and at the end we must avoid being left without strength and must avoid even more being oversatiated".[1]

In the stoic tradition, too, (e.g. Epictetus) negative emotions concerning death were viewed as senseless, for if death has taken place there is nothing we can do about it. Epictetus considered it foolish to wish to live eternally, for man should do only that which he is able to do.

Seneca and especially Marcus Aurelius stressed the need to view the inevitability of death with resignation and called for showing neither fear nor repulsion nor contempt since death is simply one of Nature's functions.

The origins of Christianity have been strongly influenced by elements of stoicism, even though a leading role within it was played by religious-philosophical conceptions concerning man's death and his soul's "immortality" that were drawn from the ancient Middle East. Christianity lent a new meaning to the rationalistic philosophical approach to death: it responded primarily not to the tragic nature of death itself but by offering a "consolation" to those who suffered from the burdens of earthly life by referring to the "miracle of resurrection" as a "transition" to a new and more happy life. The corresponding "conditions" (reflected, in particular, in the Ten Commandments) stressed precisely the moral-humanistic aspects of human consciousness and behaviour of individuals remembering death and "the day of judgement". Subsequently rationalistic revisions of that original idea did not alter its basic content, which has not ceased to greatly influence not only the undeveloped consciousness of large masses but also the moral views of many individuals who do not believe in the "miracle of Christ's resurrection".

Beginning with the Renaissance rationalistic tendencies of humanism developed within these ideas that substantially undermine Christianity's Orthodox dogmas concerning man's death and immortality and the miracle of his "resurrection". Above all, these dogmas were greatly damaged by materialist conceptions and the development of natural scientific knowledge, especially biology. They seemed to be felt to a lesser extent, however, in the moral-humanistic sphere. In particular, Spinoza considered that "a free man ... (is not led by the fear of death ...

[1] Marcus, Tullius, Cicero, *op. cit.*, pp. 384-385.

he) thinks of nothing less than of death, but his wisdom is a meditation of life".[1] It is true that in continuing to reflect on death and immortality he adopted the idea that death is a kind of eternal life and a return to a single immortal substance.

France's eighteenth century materialists offered a more consistent solution to that problem. Yet, while their ideas did undermine Christian teaching concerning man's death and immortality their ideas contained few positive approaches to a consideration of that problem's spiritual-moral aspects.

In Russian philosophy a specific approach was offered in some detail by A. N. Radishchev in a treatise *On Man, His Mortality and Immortality*. Above all, he was guided by moral-humanistic motivations: the treatise begins with an expression of hope that "after his death" the author will be able to see his friends to say "I continue to love you as before!" Examining man at the physical level, as a materialist he does not find any evidence pointing to "probable immortality", although "a desire for eternity is indeed present in man together with all his other desires"[2]. This is why, abandoning materialism, Radishchev writes that death is nothing but "a natural change in the human condition"[3], but that man's immortal "soul or what we call an intelligent being" and what "is a property of an artfully constructed body, will never be destroyed and will never vanish"[4]. In his view, immortality is an attribute of an individual's personality of his unique Self. Addressing himself to man, he concludes: "Follow the path defined by nature and have faith that if you will live beyond the end of your days and a destruction of thought will not be your lot, your future state will be commensurate with your existence.. ."[5]

Radishchev's rationalistic approach to the problem of man's death and immortality affirmed ideas that are alien to religious consciousness. Yet it contained deep contradictions that greatly weakened the materialist basis of his approach to that problem.

---

[1] Benedict Spinoza, *Ethics*, Dent & Sons, London, 1925, p. 187.

[2] A. N. Radishchev, *Izbranniye filosofskiye i obschestvenno-politicheskiye proizvedeniya* (Selected Philosophical and Socio-Political Works), Moscow, 1952, p. 329.

[3] *Ibid.*, p. 363.

[4] *Ibid.*, pp. 369-370.

[5] *Ibid.*, p. 416.

He also left "open" to analysis its moral and humanistic aspects which he basically described in terms of a tragic disharmony between what is subjectively desirable and what is objectively real. That "open" theme was discussed with particular intensity by representatives of idealist systems. This is especially true of systems of the mystical and irrationalist type, since rationalist branches of philosophical thought, represented, in particular, by the teachings of Fichte and Hegel simply "sublated" the problem of the finiteness of man's existence into a separate problem. In many cases a basis for such an approach was provided by the ideas of pantheism, which were expressed by Goethe in a whimsical as well as philosophical and poetic form, for instance, in his well-known essay *The Nature.* That approach rested on a different philosophical basis in the case of Schelling's teachings concerning the "world soul" and nature's universal spirituality.

Kant used as a point of departure the ideas of transcendentalism and devoted particular attention to the finiteness of man's individual existence. In particular, in a treatise on *The End of All That Exists* he refers to the expression concerning a dying person, "He is passing away from time into eternity", and observes that the expression looses its meaning if by eternity we imply infinite time. He stressed that "in such a case man never leaves the domain of time and only shifts from one time into another. Accordingly, one must have in mind the end of all kinds of time, and while the duration of man's existence is continuous, that duration (if man's existence is seen as a value) is conceived as a value that is altogether incommensurable with time (duratio noumenon), and we can only acquire a negative conception of it".[1] Kant noted that "such a thought contains something frightening that brings us closer to the edge of an abyss from which those who sink into it can never return;. . . and at the same time we are attracted by it, for we do not have the power to turn our frightened eyes away... It is monstrously *lofty*; partly because of the haze that envelops it, which allows our imagination to be more active than in the light of day. Finally, it is also interwoven in a surprising way with man's everyday intelligence, and this is why one finds it in some form in all

---

[1] Immanuel Kant, *Gesammelte Schriften,* Vol. III, Berlin, Georg Reiner Verlag, 1912, p. 327.

times and among all peoples who enter the path of reflection."[1] According to Kant, "the very idea of the end of everything that exists originates in reflections not on the physical but on the moral aspect of the matter, and moreover, only the latter can be related to the idea of the supersensory that is comprehended in the sphere of morality, and, hence, to the idea of eternity. Accordingly, it is wise to behave as if it were certain that another life awaits us, and that upon entering it the moral state with which we will end our present life will be taken into account."[2]

The subsequent development of thought concerning the moral aspect of the problem of man's death and immortality, directly or indirectly opposed to the Orthodox religious approach, was given an increasing emphasis even though it continued to remain in the background in the materialist philosophical conceptions of the 19th century. This is also true of positivist conceptions containing certain materialist ideas as, for example, the teachings of Herbert Spencer. He sought to resolve that problem with the help of a biologically one-sided interpretation of the conception of progress as a process of adaptation and of contribution to "general welfare".

Irrationalist conceptions of man's life and death stood in sharp contrast to these ideas. They were clearly expressed in the works of Arthur Schopenhauer and Karl Hartmann and especially of Friedrich Nietzsche. Their central theme was a pessimistic assertion that life "is an endless repetition" of something that it would be better not to have at all, that its aim is not happiness but suffering, and that death, accordingly, is its principal truth at least for a man who is able to anticipate it and expect it. In denying the resurrection of the body and the soul's immortality Schopenhauer affirmed the "eternal principle" of the immortality of the species, being realized through an irrational life principle that he called the will to live. Hartmann and Nietzsche preached the need to resign oneself to the senselessness of life and to view death as a road to what is "an eternal return of the same thing".

---

[1] *Ibid.*

[2] *Ibid.* As La Rochefoucauld wittily observed, "we fear everything as mortals should and desire everything as if we were awarded immortality".

Subsequently, the view that life's senselessness is in fact the principal meaning of life and death has also found reflection and further development in existentialism, and especially in the works of Jean-Paul Sartre and Albert Camus.

Attempts to free oneself from this sphere of either rationalistically or else irrationalistically interpreted abstract philosophical approaches to the questions of man's life, death and immortality were linked to a shift in emphasis on man as an individual and a personality constituting an integral unity of his biological and spiritual qualities. It is man's biological qualities, moreover, that acquire a dominant role. These tendencies are especially evident in the philosophy of Ludwig Feuerbach in which thoughts relating to man's death and immortality were largely confined to the purely anthropological sphere. For him individuality in a physical (physiological) sense becomes the principal object of philosophy and of a "new religion", and, accordingly, he resolved that problem equally definitely and one-sidedly. "Instead of immortal individuals the 'new religion' requires both spiritually and physically healthy persons. It attaches greater importance to health than to immortality."[1] This is also stated even more strongly: "It is true that for women to believe in immortality is womanish, while for men it is a womanish belief."[2]

Only love, in Feuerbach's view, will resolve the mystery of immortality, even though in principle, "the future should never be the direct object of one's thinking and concern. An intelligent enjoyment of the present is the only intelligent form of concern for the future."[3]

Of course, such an approach which abstracts from the spiritual and moral definition of the meaning of human life does not establish the essence of man's death an immortality. It also partly weakens Feuerbach's philosophical critique of Christian dogmas which he made in *The Question of Immortality from the Point of View of Anthropology.* In that work Feuerbach is primarily concerned with the view that belief in immortality is supposedly inherent in "human nature". He seeks to show that in fact it is not at all a direct expression of human nature but

---

[1] Ludwig Feuerbach's *Sämmtliche Werke,* Vol. 2, Druck and Verlag von Otto Wigand, Leipzig, 1846, p. 389.

[2] *Ibid.,* p. 396.

[3] *Ibid.,* pp. 396, 397.

originates in reflections and is, moreover, the product of false conclusions. For a living person a dead person is not transformed into anything at all. "In losing his bodily existence a person does not lose his existence in the spirit, memories, and hearts of living persons", and it is this that produces in "uneducated persons" the belief in immortality on which religion, too, is based.[1]

The imaginary faith of peoples in "another life" is simply, in Feuerbach's opinion, a reflection of their belief in our present life. Being satisfied with this world man cannot imagine its opposite and does not understand that "death is its rigorously inevitable consequence".[2] The situation regarding immortality is similar to the one relating to the concept of God, which man imagines in terms of his own essence.

A subjective need for belief in immortality is based on the conception of an uninterrupted continuation of the present life as infinite, which is connected, according to Feuerbach, not with a striving for self-perfection but rather for self-preservation. He arrives here at an understanding of the social significance of the question of life and immortality and considers that "immortality is essentially no more than an occupation for dreamers and loafers. An active person who is busy with the objects of human life has no time to think of death and, hence, does not feel a need for immortality. Even if he does think of death he only perceives it as a reminder to make wise use of the life capital that has fallen to his lot, not to waste valuable time on meaningless tasks but to spend it on completing his own task which he has set himself."[3]

The determining factor, accordingly, concerns the nature of the tasks that are envisaged. Feuerbach formulates them only in general terms. He turns to an individual's wordly tasks, but interprets them very abstractly. He exclaims: "Enjoy everything that is good in life and to the extent that this is in your power reduce its troubles! Believe that on earth things can be better than they are today and then they will indeed become better. Expect the best not from death but from your own selves! It

---

[1] Ludwig Feuerbach, *op. cit.*, Vol. 3, 1847, p. 264.
[2] *Ibid.*, p. 276.
[3] *Ibid.*, pp. 286-287.

is not death that you should seek to banish from this world; but troubles—those troubles that can be banished, those that originate only in the laziness, meanness and ignorance of individuals; these are the most terrible troubles. Natural death, that death which is the result of completed development of life is not a trouble; the death that results from need, vice, crime, ignorance and rudeness is, of course, an evil. That death you should, of course, seek to banish from this world or at any rate you should seek to limit its sphere of action as much as possible."[1]

These words clearly express Feuerbach's own credo. He believes that that is precisely "how reason addresses man,"[2] in contrast to Christianity which, while seeking to overcome an imaginary evil, leaves untouched life's true troubles. Yet, Feuerbach lends that thought purely anthropological form, even though he does emphasize man's special mission deriving from his property as a "moral being, that is, as a social, civic, political being".[3] He rejects the reproach that his views of death and immortality are absolutely negative and that they destroy the personality. But his conception's principal shortcoming concerns his very interpretation of personality and individuality independently of their social linkages.

That shortcoming is generally inherent in all anthropologism. While Feuerbach's anthropologism was directed against conceptions of death and immortality proposed by Christianity and also by a number of critical and rationalistic trends of idealist philosophy, the subsequent idealist philosophical anthropology (e.g., M. Scheler) and especially its existentialist variant (Martin Heidegger, Karl Jaspers, Jean-Paul Sartre, Albert Camus) addresses itself to speculative conceptions of "genuine" and "non-genuine" forms of existence in confronting death, and concerning the absurdity of death. This is also true of the interpretations of death's connection with Eros, which continue Nietzsche's line of approach. They have been further stimulated by a type of Freudianism, that is artificially "combined" with Marxism" (e. g., Herbert Marcuse, Eric, Fromm).

In Marxism it is precisely a scientific understanding of the

---

[1] Ludwig Feuerbach, *op. cit.*, p. 287.
[2] *Ibid.*, pp. 287-288.
[3] *Ibid.*, p. 335.

individual's and personality's social essence and of the dialectics of individual and historical components as they develop that makes it possible to fully overcome anthropologism in questions relating to man's death and immortality. Marxism does not remove that philosophical problem as is done by the systems of objective idealism and in diverse variants of pantheistic conceptions. It shifts them away from the sphere of man's strictly individual and natural-biological existence into a dialectical integral sphere, where man's life is combined with the human race's species life which is defined by social and spiritual factors including moral-humanistic ones. This is why Marx considered that "*death* seems to be a harsh victory of the species over the *particular* individual and to contradict their unity. But the particular individual is only a *particular species-being,* and as such mortal."[1] This is also why Marx interpreted communism as "a *genuine* resolution of the conflict between man and nature and between man and man—the true resolution of the strife between existence and essence, between objectification and self-confirmation, between freedom and necessity, between the individual and the species. Communism is the riddle of history solved..."[2]

But one of the most intriguing and difficult riddles of history is precisely the problem of death and immortality which appears as a contradiction between the individual, the person and the human race and society in their definite temporal duration which constitutes history itself. Naturally, it is not sufficient to approach that problem exclusively from the point of view of the species, and define the position of Marxist philosophy as one that is thoughtlessly optimistic on the ground that according to the Marxist view, even though the individual does die he, nevertheless, acquires immortality through the human species, his progeny,[3] and the human race's creative legacy, its material and spiritual culture.

---

[1] Karl Marx, "Economic and Philosophic Manuscripts of 1844", *op. cit.,* p. 299.

[2] *Ibid.,* pp. 296-97.

[3] Denis Diderot observed that "for philosophers the role of progeny is similar to that of the other world for believers". (Denis Diderot, *Correspondence,* Les Editions de Minuit, Paris, 1961, p. 67.)

There is also the individual aspect of the problem of death and immortality, which is keenly felt by the personality. It contains a tragic element that no philosophy, however, optimistic, can overcome. Let us recall the words of Alexander Pushkin:

The time has come my friend! For peace my heart is yearning
As days succeed each other each one carries away
A particle of life, but death may come at once
Just as we make our plans for a long life together.[1]

And yet:

To mortal hearts all threats to life
Bring a mysterious delight.
A promise, it may be, immortality,
And happy are the few who in the midst of cares
This knowledge felt and knew.[2]

And yet again this is followed by different thoughts and sentiments that at first glance appear to be contradictory:

I do not want, O friends, to die:
I want to live, and think, and suffer.

That profound combination of a conception of life as spiritual happiness of thinking with that of suffering has been noted by many authors. In the present context, however, we are interested in another aspect, namely, that it is both necessary and just that individuals should die in order to give way to others so that life be eternal. To some extent an awareness that while the individual dies the personality does not die completely but lives

---

[1] This same thought is also brilliantly expressed by Anna Akhmatova:

But when the music began to sound
And winter awoke all around,
Everyone could see Queen Death herself
As she stood beside the pier.

[2] Akhmatova expressed this in poetic form in the following words: "As I die, I yearn for immortality..."

in the deeds and memories of subsequent generations, just as Pushkin will always continue to live within us ("No, I will not die completely"), may reconcile us somewhat with death. Although it is clear that the life of a great personality is only a unique example for all to follow and this does not alter the tragic nature of a personal encounter with death.

This is why it is not optimism but rather *realism* and more precisely *scientific real humanism* that must serve as the moral-philosophical foundation for the Marxist approach to the question of man's life and immortality. That approach does not provide conclusive solutions applicable to all. But it does clearly represent its general world-view and approach to such solutions, which are so dissimilar and intellectually and emotionally unique for each individual. Georgi Plekhanov had already observed that "from the point of view of modern natural-scientific knowledge there is nothing mysterious in death. Death is not at all a sphinx. Shelley's words concerning the late Keats can apply to any dead person: 'He is made one with Nature.'[1] To be made 'one with Nature' does not constitute anything mysterious, but it does often bring much pain especially to those who have lost a close person."[2] This is precisely the circumstance which quite aside from the point that such persons themselves no longer fear the thought of death, casts light on the problem of *one's own* death in the context of other socio-ethical and moral-humanistic aspects.

It seems to me that it is extremely fruitful to approach the problem of death not simply "from the viewpoint of natural science" supplemented with emotional experiences, or from a purely moral-philosophical position, in an attempt to "overcome" these emotional experiences, but in the context of an *integral unity* of both. Moreover, this should be done with due regard not only for personal emotional experiences and reflections concerning *one's own* death as a phenomenon relating specifically to oneself, but also for the way in which it is reflected in others, for whom one's life is a pure reflection of awareness and emotions[3]. In this respect one can learn much by compar-

[1] Plekhanov asked that these words be engraved on his tombstone.

[2] G. V. Plekhanov, *Sochineniya* (Works), Vol. 14, p. 90.

[3] On a philosophical plane the substantial difference that exists between one's own death and that of other was stressed by Heidegger and

ing the reflections of natural scientists on death with those of humanist literary writers. I have already made such a comparison in the case of Mechnikov and Lev Tolstoy and here is another one concerning the views of Vernadsky, an outstanding Russian natural scientist, and, in part, of Lev Tolstoy, but especially of Prishvin, a prominent humanist writer.

Unlike Mechnikov and subsequently Shmalgauzen, Vernadsky did not explicitly consider the problem of death and immortality as a natural scientist. Nevertheless, he was interested in that problem not merely as a mortal and thinking person, but also as a scientist seeking to define his own position in life and science. Already in his youth he wrote in a diary (15 July, 1884), that is, long before Mechnikov published his first works, but at a time when Lev Tolstoy had already begun to widely preach his ideas, that death would come and "after that I will cease to exist, and there will be nothing. I do not know whether such a feeling is frightening. One can perhaps better describe it as strange. Death seems frightening only when you compare subsequent inactivity and cessation of existence to current living personalities. To them death appears as something frightening, but since after death they will no longer exist, then the fear of death, too, should not exist for them... Since I do not recognize either a soul or life beyond the grave, I do not see that some kind of loss takes place at this point."[1]

In rejecting "well-known consolations provided by faith", Vernadsky clearly defined the meaning and aim of life proceeding from the premise that "everything lies in this life, and not in another or a hundred others that are said to follow. And it is in this life that one should strive to achieve impossible happiness."[2] For Vernadsky this is, first, the development of science, and secondly, "the development of the human race". In both cases, moreover, Vernadsky also viewed the achievement of these objectives as a personal pleasure. He understood the need for

---

Sartre with references to Lev Tolstoy, who noted in *The Death of Ivan Ilyich* that "...the very fact that a close acquaintance had died produced, as always, among all those who knew him, a feeling of joy that someone else had died."

[1] *Stranitsy avtobiografii V. I. Vernadskogo* (Pages from V. I. Vernadsky's Autobiography), Moscow, 1981, p. 46.

[2] *Ibid.*, p. 133.

action and active love for science and for mankind, "a boundless love that makes it easy to sacrifice everything, oneself, absolutely everything. And for those persons who do not hope to find a better life beyond the grave ... such a feeling and the support that it provides is necessary, without it life is not possible for them. . ."[1]

His serious reflections concerning the meaning of life also produced a love for philosophy. In particular, Vernadsky studied the works of Marcus Aurelius, and in his youth found much in stoicism for "man's purely moral and mental development".[2] A decisive role, however, was played by Vernadsky's conscious, i.e., philosophically expressed materialistic approach that rejected all religious-idealistic forms of world contemplation. In his own case that approach often assumed highly original forms. For example, in criticizing pantheism he stressed ideas concerning the role of a creative element in man's own consciousness or spirit. "I cannot understand pantheism, just as I cannot understand the joy and purpose of merging with Nature, nor the satisfaction that is provided by an awareness that one's own death (and in this case the whole life essentially reduces to an ability to die well) is useful for a larger cause. I can also only conceive of a single concept of religion, one in which there is an eternally existing, immortal person, and I view all divinities as secondary phenomena in man's religious feeling."[3]

Naturally, in this respect Vernadsky differed substantially from Tolstoy, although possibly not as sharply as Mechnikov, since he touched less frequently on moral-philosophical problems that were "sensitive" for Tolstoy. He recalls that in his youth Tolstoy's *Confessions* left a deep impression on him, while his *Gospel* "left me cold" even though it did force him to "think about many matters"[4]. In reading Tolstoy's later works, including his *Tales of Sevastopol, War and Peace,* and *The Kreuzer Sonata,* Vernadsky noted, "I see before me his entire personality, whose constancy I find surprising: . . . this is still the same man, sensitive, strong. His basic trait is a suffering resulting from

[1] *Ibid.,* p. 35.
[2] *Ibid.,* p. 63.
[3] *Ibid.,* p. 133.
[4] *Ibid.,* p. 35.

a lack of a deep faith in the integral character of life..."[1] These differences in their views were also reflected during a meeting between Vernadsky and Tolstoy that he subsequently described in his diary as follows (29 April, 1893): "... L. N. Tolstoy visited us and we had a prolonged discussion about ideas, science, etc. He mentioned that he is viewed as a mystic, but it is rather I who is a mystic. And I would gladly be one were it not for my skepticism. I think that Tolstoy's teaching contains much more that is profound than I had initially imagined. And this concerns: 1) life's basis is a search for truth; and 2) the real problem is to express that truth without any concessions. I believe that the latter is the most important thing, and that a rejection of all hypocrisy constitutes a teaching's main strength, since it is then that a personality expresses itself most strongly and acquires social power. Tolstoy is an anarchist. He values science and the search for truth—but not universities, etc."[2]

We see that unlike the content of Tolstoy's conversations with Mechnikov, in which emphasis was placed on issues relating to man's life, death and immortality, his discussions with Vernadsky primarily concerned science. Their views on questions that mattered most for Tolstoy were so different that they chose simply not to touch on them.[3] Nor did Vernadsky hide this in a

---

[1] *Stranitsy avtobiografii V. I. Vernadskogo, op. cit.*, p. 133.

[2] *Ibid.*, p. 129.

[3] Tolstoy also displayed tractability so uncharacteristic of him with regard to the scientific views of N. N. Miklukho-Maklai. In a letter to that remarkable Russian scientist and traveller he writes that after reading his brochures "I found in them certain things that interest me. No, not interest but touch and delight me, since you appear to be the first person who has shown beyond all doubt through experience that man is man everywhere, i.e., a kind, sociable being, with whom one may and must communicate only through kindness and truth rather than through cannon and vodka." And Miklukho-Maklai achieved this, Tolstoy writes, "under the pretext of scientific activities". It is true that following these words Tolstoy asks to be forgiven for a "frank expression of my convictions". And later the secret of that politeness is made clear. "I do not know what contribution to the science that you are serving will be made by your collections and discoveries, but your experience in communicating with savages will mark an epoch in the science that I myself serve, the science showing how people should live with each other. If you will write that story this will be a great and important service to mankind." Continuing his letter Tolstoy asks that Miklukho-Maklai

letter to Tolstoy (August 9, 1901) in connection with the latter's illness (written also on behalf of his wife): "While in many respects we hold views and opinions that differ from your own, your striving to express truth as you understand it did not pass unnoticed in our spiritual life..."[1] Many years later (September 9, 1941), recalling his conversation with Tolstoy, Vernadsky noted that "at one time I believed in personal immortality... But already long ago, especially after I began to see life as a planetary cosmic phenomenon and after creating the concept of living matter, i.e., after 1916, the question of personal immortality lost its meaning for me as a scientific question."[2]

"It became clear to me that: 1) man as a personality is also a planetary phenomenon (the noosphere); 2) that everything in the world, the cosmos, is based on time and space-time; 3) that there is only one scientific way to arrive at truth and this is the cognition of truth trough collective generations rather than a philosophical and religious search for truth."[3] This represents Vernadsky's own summary, as it were, of his prolonged "inner debate" with Tolstoy, which led him to provide his own positive answers to its questions.

In spite of these contradictions in the way in which prominent Russian scientists and humanist writers perceived the problems of man's death and immortality they show a clearly delineated and profoundly *common* humanistic line of development that is associated with the specific social conditions and spiritual situa-

---

should not object "to my attacks on scientific observations. I take these words back..." Miklukho-Maklai received his advice with such gratitude that he even assured Tolstoy that he would not object to his "attacks on the science for which I have worked all my life and for which I am always ready to sacrifice everything". (*Chelovek s Luny. Dnevniki, Statyi, Pisma N. N. Miklukho-Maklaia* /A Man From the Moon. Diaries, Articles and Letters of N. N. Miklukho-Maklai/, Moscow, 1983, pp. 4-6.)

[1] Vernadsky, *op. cit.*, p. 183.

[2] In effect this same thought was also expressed by Ernest Rutherford, a prominent physicist, in his autobiography. In it, while saying nothing to either confirm or deny the doctrine of immortality, he stated that individuals could be happy without it even in difficult times. In his opinion, to ascribe all our actions to a search for immortality is an extreme form of madness that produces a life-long hope and fascination, but death then arrives before we have had time to enjoy a single hour.

[3] Vernadsky, *op. cit.*, pp. 145-146.

tion within which Russian culture developed. While to a certain extent Vernadsky's position and subsequently that of Shmalgauzen differed from the ideas of Mechnikov's "optimistic philosophy", they, nevertheless, had much in common and also shared many ideas with Tolstoy, who dissociated himself from "scientific science" so emphatically.

The foregoing is also true of the views of Prishvin, which are marked, on the one hand, by a closeness to the "natural-philosophical pantheism" of natural scientists, and on the other, by a clear emphasis on the moral-humanistic side of the problem of life and immortality in the manner of Tolstoy. In his diaries (which he viewed as his life's principal work, written "in response to the needs of a growth in consciousness and to that alone. . ."[1]) Prishvin bases his argument on a proposition that "man's health lies not in his heart or his kidneys, not in his roots, his leaves or his spine" (cf. Tolstoy's similar words) but in the good that he can say or do to another person: "it would even appear that this is a law: if I am happy then everyone must be happy".[2] And even the loss of what is external in human health frequently results in something that is better: an elderly person neither regrets the past nor envies the young, for "I have a friend, I love—therefore I am".[3]

He also said brilliantly about those "who feel the poetry of everyday life's fleeting instants and who suffer from their inability to grasp them", "the precious nature of human life", the thought that in human creativity the "Self" must "look at itself in the mirror of eternity, and always be victorious over flying time".[4] Prishvin feels acutely his oneness with Nature, but unlike Shelley, not only in death but also in life. This is why "art and science—are doors, as it were, from the world of Nature into the human world: Nature enters into the world of man through the door of science, and man enters Nature through the door of art and then recognizes himself and calls Nature his mother".[5] Such a "humanization" of Nature is so intense that

---

[1] M. M. Prishvin, *Doroga k drugu. Dnevniki* (The Road to a Friend, Diaries), Leningrad, 1982, p. 11.

[2] *Ibid.,* p. 6.

[3] *Ibid.,* p. 9.

[4] *Ibid.,* p. 13.

[5] *Ibid.,* pp. 34-35.

for Prishvin "to feel Nature is to feel one's personal life reflected in Nature: Nature is then myself".[1] This also leaves its special imprint on his moral-philosophical perception of man's death and immortality: "though man may die, even his fragments will carry his victorious effort on the road to immortality".[2]

Prishvin acutely experiences that problem, connecting man's activity, love, kindness, striving for turth, etc.—with the way he serves others: "Man lives and creates something new, and there then remains forever something unique that he creates through his work, deeds, thoughts, his greetings, clasping another person's hand, or simply through a smile."[3] Here one sees a strong connection of Prishvin's moral-philosophical ideas with those of Tolstoy. In a note in his diary Prishvin observes: "An individual cannot define personal happiness as his objective. Happiness is awarded as a gift to those who set for themselves and achieve life's most difficult objective, namely, the happiness of a close person."[4] In fact in his diary (September 15, 1889) he referred directly to Tolstoy's widely known statement to the effect that "the task of life and its mission is joy", which, according to Prishvin, "serves to prolong life"... And it follows that "if one day I will die, someone else will enjoy the life at this lake with swans".[5]

In emphasizing the importance of the integral personality, "unique, unrepeatable and necessary" to people "as they move upwards towards the skies and spread along the earth",[6] Prishvin concludes his diaries with a wise pronouncement: "Before dying I would like to renounce myself in order to assert the joy of life: I cannot, but you, friends, continue to reign." He understood very well that "the ethics of socialism lies in giving smaller persons the souls of great individuals".[7] For only if man possesses an "integral personality" will he be able also to solve complex moral-philosophical problems of death and immortality with wisdom and in accordance with the principles of scientific, real humanism.

[1] *Ibid.*, p. 41.
[2] *Ibid.*, p. 52.
[3] *Ibid.*, p. 121.
[4] *Ibid.*, p. 132.
[5] *Ibid.*, pp. 21, 23.
[6] *Ibid.*, p. 140.
[7] *Ibid.*, p. 156.

3. The Idea of Man's Life, Death and Immortality in the Light of Modern Science: Advances of Reanimatology and Myths about "Life after Life"; Humanistic Problems of Thanatology: "the Right to Die", and "the Culture of Dying"; Man's Uniqueness and the Scientific Search for Extraterrestrial Forms of Life and Intelligence; the Immortality of Man's Mind and Humaneness—the Immortality of the Human Race

In our day, the scientific, social and moral-humanistic aspects of the problem of man's life, death and immortality are attracting increasing attention not only because of the increasingly realized exacerbated personal alternatives of being but also in connection with advances in biomedical research. This has already been noted. We are referring not to some vague distant prospect of gene and medical technology, but to existing achievements that are already presenting the human race's "eternal" problems in ways that are in many respects new and that call for new definitions of a number of humanistic principles that serve to alter not only the ethics of life but also that of man's death.

Already now many scientists are raising the problem of complementing biology, the science of life, with new ideas relating to the biology of death. The questions that arise define a multiplicity of highly controversial moral-humanistic problems that frequently defy traditional approaches. This is true, in particular, of modern reanimatology which has achieved substantial successes in restoring to life individuals who were in a state of clinical death.[1] Here is one such problem. It has been established

---

[1] R. A. Mody, *Life After Life. Investigation of a Phenomenon: Survival of Bodily Death,* New York, 1978; *Life After Death.* Contributions by Arnold Toynbee, Arthur Koestler and others, London, Weidenfeld and Nicolson, 1977.

See also V. A. Negovsky, "Nekotoriye metodologicheskiye problemy sovremennoi reanimatologii" (Selected Methodological Problems of Modern Reanimatology), *Voprosy filosofii,* No. 8, 1978. V. A. Negovsky scientifically exposes shortcomings of various myths concerning "life after death" that are being resurrected in connection with advances in reanimatology, and which themselves seek to "reanimate" the conception of an "immortal soul" (V. A. Negovsky, "Ob odnoi idealisticheskoi kontseptsii klinicheskoi smerti" (Concerning an Idealistic Conception of Clinical Death) in *Filosofskie Nauki* (Philosophical Sciences), No. 4, 1981; also V. A. Negovsky, *Vtoraia zhizn* (Second Life), Moscow, 1983.

that when an organism dies the brain's cortex with which man's conscious activity is associated perishes first, while its heart can continue to work the longest. Methods of reanimation make it possible to restore the work of the heart, and hence to return a person to life. But since, as V. A. Negovsky notes, there is not a single known case of restoration of brain activity following a diagnosis of death, the individual's personality is never restored even though his organism may be returned to life. Negovsky is right in sharply posing the problem concerning how one should evaluate the "death of the brain". Since individuals then lose their personal qualities irretrievably, is it still possible to speak of their life?

Today many national and international medical organizations have expressed support for the proposition that irreversible damage to the brain, rather than the cessation of heart activity be employed to define the moment of a man's death. In such a context, Negovsky is entirely right in raising a socio-ethical and moral humanistic issue of the greatest importance. "False humanism prescribes that medical doctors initiate reanimation measures and continue them so long as breathing and heart activity has not ceased. And yet world-wide statistics show beyond doubt that if under normal temperature conditions and in the absence of anesthetics blood circulation ceases for a period that exceeds the brain's capacity to withstand this, then the restoration of the functions of the central nervous system, which is needed to preserve the patient's personality, ceases to be possible. We are inclined to agree with the view that it is inappropriate to initiate the restoration of heart activity and of breathing if it has been reliably established that the cessation of heart beating exceeded the maximal period that the brain could subsequently survive. Once the presence of irreversible damage to the brain has been established, reanimation measures should be terminated."[1]

Those are the words of a prominent specialist, and therefore the way he poses questions of a moral-humanistic nature is given some practical elucidation. It must be said that the discussions of practical issues associated with man's death, in which the

---

[1] V. A. Negovsky, "Nekotoriye metodologicheskiye problemy sovremenoi reanimatologii" (Selected Methodological Problems of Modern Reanimatology), *Voprosy filosofii,* No. 8, 1978, p. 71.

participation of philosophers, ethicists, sociologists and other specialists is still insufficient, sometimes acquire highly unexpected forms. And in a number of cases they already produced questionable and sometimes simply false and even anti-humane conclusions. For example, a comprehensive survey prepared by the United States National Academy of Sciences' Commission on Science entitled *Biology and the Future of Man* states that today the problem is being posed of finding ways to alleviate tragic situations in which bodily life is preserved while the mental possibilities that characterize a normal personality are not.[1] This draws further attention to issues that can no longer be postponed. Is a society justified in continuing to preserve the lives of elderly persons who have lost those mental functions that distinguish human beings from other living bodies? Where is the limit beyond which one cannot preserve the contacts of these old persons with society? As the authors of the survey believe, these are increasingly urgent questions. A highly developed civilization must solve them intelligently. In the meantime the need to support research activities that are intended to help preserve the health of newly born children and young people generally compete for human and material resources with the need to study questions of aging. Tendencies to give greater support to children and young people may limit those efforts of societies that might keep elderly persons active beyond reasonable limits. Similar considerations must also be applied in planning resources for the care of non-productive persons by comparison with those who are potentially productive. Virus child diseases and hereditary illnesses, child leukemia, diseases of the endocrine system, traumas, accidents, etc.—all seem to call for greater attention than efforts to prolong life. But that, the authors conclude, are cruel decisions for which no absolute methods exist. A nation that possesses the needed resources may afford attacks on all fronts. It is also noted that generally research activities cannot rest on compulsion.

Naturally, these considerations of the survey's authors contain much that is controversial and that requires both analysis and criticism. It must be said that there is currently much discussion of the philosophical aspects of the problem of human death,

---

[1] *Biology and the Future of Man*, New York, London, Toronto, 1970.

and in such a context particular interest attaches to the issue of the "right to die" with regard to which two opposite positions have been clashing. One affirms an unlimited freedom of individuals in deciding on such questions, while the other subordinates it to the interests of society and of the state (the conception of so-called paternalism). In part, this has also been reflected in the documents of the World Health Organization to which I have already referred. There is a sense in which the term "right to die" itself sounds paradoxical: for over the ages all human rights were based on man's most important right, namely, the right to live. Any of the rights that have ever been proclaimed may be viewed as a further development, widening and concretization of that fundamental right, for any one of the subjective rights is necessarily a specific expression of the right to live, the satisfaction of one of man's vital needs, interests and aspirations. In our own time, however, as a result of the intensive development of medicine, the question of life and death has emerged as a problem of choice. That choice, moreover, is effected not only by the individuals whose life and death are being considered but by other persons as well. In situations in which death is an outcome that partly lends itself to control "the right to die" raises a number of problems: is the right to live not only a right but also a duty and obligation, and must society protect a person's life independently of his own will? While similar questions arise in the case of suicide, today's discussions concerning the "right to die" refer not to suicide as an act of an active subject, but to that of a dying individual as a passive object for whom death is artificially delayed. In such a context the problems of euthanasia[1] and artificial methods to prolong life become central in such discussions. Modern philosophers, lawyers and medical doctors are emphasising the two most fundamental issues: assuming that euthanasia can be morally justified, under what conditions should it be made legitimate? Most scientists adopt an anti-paternalistic position and consider that the right of free choice is the most important moral principle which should be embodied into law in so far that this is possible, for limiting the freedom of an

[1] Euthanasia (from the Greek *euthanasia*) refers to painless, quiet, "serene" death (especially in hopeless cases).

individual is morally unjustified whenever his own actions do not harm others. . . Should a patient decide to hasten his own death, without causing harm to others then the act of euthanasia, anti-paternalists affirm, is perhaps an expression of individual freedom and should not be prohibited by law. The corresponding arguments of anti-paternalists run as follows: modern medical technology continues to substantially increase possibilities for prolonging life; sometimes dying persons themselves notice a gradual destruction of their organism, and a fading of all forms of activity; such persons are not only subjected to continual physical suffering but are aware of the burden that they constitute for others, and for persons who are close to them. In such cases it is "amoral" not to allow a person to die.

Some scientists who are inclined to accept paternalism object to euthanasia by arguing against a moral justification for depriving a person of life as follows. First, human life is sacred and, accordingly, euthanasia must not be applied under any circumstances. Such a view is either based on religious grounds, or else expresses the conviction that the sacredness of human life forms the very basis of social order. Secondly, euthanasia lends itself to possible abuses by doctors, family members or other interested persons. And finally, euthanasia contradicts the principle that "so long as there is life there is hope" and neglects the possibility of a diagnostic error. In such a case the consequences of euthanasia are catastrophic. Moreover, after the death of a patient to whom euthanasia was applied a new remedy may appear capable of curing cases which were formerly incurable.

Many authors turn to philosophical definitions of life in seeking to define, in concrete terms, when a person actually dies, and doctors have the right to disconnect artificial life-support systems, i.e., apply so-called "passive" euthanasia. Two basic points of view are discussed. According to the first, a person's life should be protected to the very last moment, while according to the second view, it is possible to pronounce a state of death and disconnect support systems following the cessation of activities in the brain's cortex.

It has already been noted, however, that this alternative, namely, full personal autonomy or absolute paternalism, is based on false methodological premises that do not consider the dialectics of the mutual relations between the individual and society, which defines the meaning of individual human lives depending on which the question of a person's "right to die" may also be settled.

Quite fruitful are attempts at an interdisciplinary analysis of that problem that also includes the moral-philosophical aspects of the question of death. This is clearly evident, in my opinion, in Warren Schibles' interesting study.[1] Of course, there are other solutions to that problem,[2] but I will limit my analysis to Schibles' point of view, which is highly remarkable. While he does this in a somewhat exaggerated manner, he correctly notes that among the greatest questions that man has ever been confronted with, the questions of "what is death?" and "what can we do about it?" are in a special category. Until now individuals have relied on an unusual approach, namely, they either simply ignored them, or else associated their solution with various superstitions and mystical ideas. Death is a type of forbidden theme and this is why there have in effect been no honest and explicit studies designed to analyze these questions.

Many persons believe that they answer the questions associated with death by attending church services. However, modern theologians emphasize that religion does not deal with questions of specifying the concept of death and related matters. It is also widely believed that religion's aim is to show individuals how they should live and under what conditions they can live correctly. In any case, the author states, religion does not provide an adequate and objective method of research. This role, in Schibles' opinion, must be assumed by philosophy relying on the findings and methods of many sciences.

The author believes that a basic conclusion will then concern the importance of and need for a researchwise attitude as such, irrespective of the object of research. The arguments on which he bases that conclusion include the following:

---

[1] W. Schibles, *Death. An Interdisciplinary Analysis,* The Language Press, Wisconsin, 1974.

[2] These include: C. Chabanis, *La Mort, un terme ou un commencement?* Fayard, Paris, 1982. This study, which is based on the author's interviews or dialogues with a number of individuals including most prominent scientists and philosophers, explores various aspects of the problem of death. Also P. Aries, F. Dolto, F. Marty, G. Raimbault, L. Schwartzenberg, "En Face de la Mort", *Epoque,* 1983.

1) Emotions are usually viewed as a specific type of conducting channels governing our feelings. To produce positive emotions and avoid such feelings as fear, uncertainty and shock, our evaluations must be clear and intelligent—hence the need to study the phenomenon that produces particular emotions.

The study of *what* is death is the best way to struggle against the fear of death. Even simple discussions with other people as to what is death produce a therapeutic effect.

2) An awareness of death causes an individual to re-evaluate his life. In such a case a researchwise approach helps ascertain life's principal content and re-evaluate the objectives set by man.

3) A researchwise approch makes it possible for man to have a clear idea of death. The very question of the nature of death and how one may cope with it is the greatest of all challenges to the human mind. It is unworthy of man to ignore these questions, to deny them, or else construct myths about immortality. Attempts to evade an answer to these questions (whether by sitting before a television set, engaging in art, sports, collecting objects, gardening, church, money, etc.), however valuable, do not provide a solution to these questions and therefore cannot be regarded otherwise than escaping them. No philosophical system can be complete if it does not provide honest and objective answers to the questions that are associated with death. In that sense one may state that "to philosophize means to learn how one should die".

4) An ascertainment of what is death is important for each scientific discipline. But such studies are especially important in the case of philosophy and the biology of aging: their findings can help aging persons to prolong their life and make it healthy both physically and morally. We still do not know the causes of aging and this is a kind of indicator of the present level of scientific knowledge.

5) A solution to the problem of death can help man approach the very fact of death more honestly and realistically, repudiate the idea of "life beyond the grave", and abandon self-deception and the decepion of the people around him.

6) In order to understand "what is death" one must understand "what is life". And that, in turn, requires that we have a clear idea of our language, the environment, ourselves, our ethical conceptions, our psychology, science as a whole, etc.

Thus, it is not only true that "to philosophize means to learn to die" but also that "to learn to die means to philosophize".

7) A researchwise approach reduces the danger that one may become prisoner of some particular model or myth: it enables one to see prospects and possible interpretations.

8) Current research methods are often mere modifications of the primitive superstitious ideas that were shared by people throughout the centuries. In this sense, if one does feel a great need for ritual worship one should not hesitate to light a candle to Research.

9) An awareness of everyone's mortality can also cause an individual man to understand that he himself possesses an inherent desire to make that short life beautiful both for himself and for others. This should be encouraged by knowledge and research rather than by threats connected with man's inborn sinfulness for which he will be punished in hell.

10) Research is needed on questions relating to suicide (what is its cause and how it can be avoided), and must help man develop an idea of the impossibility of suicide, thus enabling him to continue the process of research.

11) Aristotle viewed the contemplating of truth as man's highest aim which makes him perfect. Referring to the immortality of the human "soul", he implicitly admitted that only research and the knowledge obtained as a result of it make man truly immortal. Man dies but knowledge remains. It is knowledge that enables man to avoid errors thus making him more free in his actions.

12) Man's basic aspirations concern understanding, sympathy, care for others, friendship, and emotional comfort. Often, however, individuals are deprived of all these because of their inability to understand their own emotions and feelings as well as of others. The author has sought to clarify this issue and he thus believes that what was begun as a study of death ends, in a sense, as a study of life and love.

In his study, Schibles refers to the data of a direct poll on questions of death. The following questions and answers provided by students are of interest. Replying to the question: "What is death?" it was stated that death is a cessation of the human body's functions, the end of existence, the end of everything, the beginning and the end at the same time, life's most important part, something that we cannot feel, a step to a higher and more intelligent form of existence, it can only be described through metaphors.

"Can you imagine your own death?" Fifty per cent of the answers were "no", including, "I do not wish to imagine this", "I am afraid of imagining it", "No, we take both life and death on trust"; "No, but I can imagine my own funeral".

"Is death a necessity, or is it a type of ailment that will eventually be overcome?" Here again answers were also almost equally divided. Some thought that death could be fully or partly eliminated; or that it cannot be vanquished, that it is necessary for the evolution of the soul, in order to determine life, or to control population. Others believed death is neither necessary nor can be overcome, it simply exists; that everything depends on how it is defined; that it does not need to be overcome.

"Is death a good thing or a bad thing? Why?" The answers were "yes", "no", "neither nor". One student answered that death is "not what I'm living for".

And still another question is interesting: "What do you think of the ceremonies that are associated with death?" Many replied that the rituals associated with funerals are "barbarous".

"Did you ever wish to die? If so, why?" Twenty-eight students answered "yes". The reasons that were mentioned included: Life appeared to be a useless waste of time; I hated my life; I was tired of living; social and personal difficulties; I wished to free others from my presence; I wanted to know what would come after; I wanted others to feel sorry for me. "Have you ever considered or else attempted to commit suicide? If so, why?" "Yes"—22 answers. "Are you afraid of death?" "Yes"—21 answers. "Do you believe in immortality?" "Yes"—15 answers. "Do you believe that you have a soul?" "Yes"—15 answers.

"Do you think that science will succeed in making man fully or partly immortal?" "Yes"—18 answers, "no"—6 answers, "perhaps"—12 answers. Explanatory comments include the following: if man should become immortal then he would cease to be a "man"; men will still find ways to destroy each other; science is a farce.

"Would you like to know the date of your own death?" "No"—26 answers, "yes"—8 answers. Some persons wanted to learn it just before their death. "If this were actually true would you like your doctor to tell you that you are fatally ill?" "Yes"—42 answers, "no"—5 answers. "Should a dying person's life be sustained under any circumstances?" "No"—30 answers. "When should the life of a dying person be sustained?" "Never"—3 answers, "Depending on the person"—8 answers, "In all cases whenever it concerns a mortal illness"—8 answers. "Were you honest in your views and evaluations of death?" "Yes"—42 answers, the others were not certain of this.

These answers, of course, are not very indicative, but they do create a certain personal-emotional background for analysis and, in particular, for a historical-philosophical consideration of the problem of death and immortality, and for analyzing the ideas of death that have developed in psychiatry, psychology and "philosophical psychology", where the author, while criticizing other points of view on the problem of death, also presents his own.

In considering these points of view Schibles emphasizes that they all associate death with negative emotions that can themselves become the cause of death (shocks, accidents). This points to the need to overcome such emotions. In the author's view, the current "therapeutic methods" (funeral rituals) simply aggravate such negative emotions rather than overcome them. In that connection he disagrees with the widely held view that man should not retain negative emotions, but should "externalize them" lest they produce various forms of mental derangement (Freudianism). Above all, he notes that there is no evidence that the human brain stores any emotions, and secondly, because negative emotions are viewed as inborn emotions for which "one is not responsible". These shortcomings follow, in the author's opinion, from the tendency of such points of view to consider individual emotions as reflections of autonomous entities.[1]

The author asks: "What is the funeral for in the first place? So that the soul will go to heaven, so that one will increase his social status, so that the emotions of the outlookers will be assuaged?"[2] In his opinion, without a clear idea of the nature of

---

[1] Schibles, *op. cit.*, p. 204.

[2] *Ibid.*, p. 207.

emotions one cannot view a particular ritual as either "good" or "appropriate". These very notions are themselves evaluative and, hence, have no autonomous meaning outside of specific contexts. In particular, a person who has never given any thought to what is death cannot acquire specific criteria for evaluating various rituals and may simply view as "best" the one that is most costly. Examining existing funeral rituals, the author comes out against various types of religious rituals, although he emphasizes that he does not oppose intelligent symbolic rituals associated with death, providing that they actually be intelligent rather than based on various types of mystical and superstitious ideas. In support of his assertion that negative emotions associated with death are not at all necessary and may be replaced by positive ones, he adduces the fact that different cultures practice different rituals. In short "grief behaviour, as other negative emotive behavior, is simply culturally imposed and learned, . . . is not necessary or inherent".[1]

In that connection the author attaches great importance to the "art of dying". In particular, he cites abundant evidence that medieval man, for example, was literally surrounded by symbols of death, gave it constant thought and prepared for it. He notes that the maxim of the ancients "remember death" *(memento mori)* was in effect the opposite of their own call to "enjoy life while you still live" *(carpe diem)*. Finally, the author notes that books concerned with the art of dying presupposed the existence of a way of dying that is "correct" or "best". But such methods have no independent meaning and acquire a meaning only in the context of specific religions.

Of course, such a categorical statement is probably unjustified, for a man deprived of a religious belief is unable to find other forms of moral support in facing death. This is an important question that should be elaborated in terms of scientific, real humanism as it relates to questions of life and death, to the art, or more precisely, the culture of dying. The philosophical basis that it provides centres neither on fantastic dreams and hopes nor on negative emotions and painful psychological tensions produced by panic, but on an honest and courageous approach born of the wisdom of personal solutions to that problem.

---

[1] Schibles, *op. cit.*, p. 209.

This is equally true of our attitudes towards the death of *others,* as well as towards society as a whole which in some way or other responds to the death of an individual personality. The relevant rules of behaviour and rituals developed by religion are increasingly losing their meaning. But what can be offered instead? And is everything here in accord with the intelligence and humaneness of individuals who are free from religion's dogmas, myths and cults? I believe that there is much moral-philosophical as well as *practical* work to be done in this area with a view to developing a *culture* in reacting on the death of others (including burial ceremonies) and not only of outstanding persons but of everyone.

These questions form an indispensable part of our life and in taking care of it we must place more emphasis on moral-philosophical and practical problems aassociated with man's death, and approach them from positions of scientific, real humanism. *In no human problem man should leave the boundaries of his own essence, which defines his intelligence and humaneness. And this is the only worthy perspective that lends meaning to his individual existence and to the historical development of the human race as a whole.*

I stress this because it is precisely in the ideal sphere of the mind and of humaneness that man's essence, which also determines his prospects, is manifested to the utmost. Lenin noted: "Man needs an ideal, but a human ideal corresponding to nature, not a supernatural ideal..."[1] Scientific philosophy and real humanism based on it provide such an ideal, which defines the meaning of human life in its individual, personal, human and social parameters. That ideal also asserts the dialectical interrelatedness of the natural-biological and the social, the finite and the infinite, man's death and immortality, and attains its most developed forms in the human race's material and spiritual culture. Scientific philosophy establishes certain ideal objectives that produce a sensation of the integral unity of life, death and immortality, of the unlimited duration, infinity and eternity of the life of each unique personality. It is this that provides a basis for the regulating role of morality both in a person's indi-

[1] V. I. Lenin. "Conspectus of Feuerbach's Book *Lectures on the Essence of Religion*", *Collected Works,* Vol. 38, p. 75.

vidual life and in his attitude towards death and immortality.

The immortality of the human race lies in the immortality of man's mind and humaneness. Above all, it is this that determines the global mission of men and of the human race, and their responsibility for preserving life and intelligence on our planet. Millennia will probably pass before the vast potentials of the mind and humaneness that are contained in man will be fully realized—and that will be his true history, the time of man's "absolute movement of becoming" (Marx) in a genuinely human, communist society.

But there is also another side to this question that concerns the human race's natural-biological infinity and the immortality of its reason, as well as the possibility of other forms of life and intelligence, of extraterrestrial civilizations in the infinite Universe. Today that extremely interesting aspect of the question is being intensively discussed in scientific and philosophical literature. The human race's breakthrough into outer space and its future emergence in the infinite expanses of the Universe also greatly alter our ideas of time, our understanding of the meaning of human life, its duration, death and immortality. This will also lead us to realize the cosmic predestination and responsibility of man whose own immortality also derives from that of his intellect and his humaneness.

Naturally, these problems call for a separate philosophical research based not only on general futurological studies but on an analysis of scientific principles that are being increasingly elaborated by scientists employing the most diverse mathematical and technical methods and means of research. And while the problem of extraterrestrial life and intelligence remains open and highly controversial, efforts to solve it are stimulating not only the scientific and technical development of human civilization but also lead us to realize the uniqueness of man himself and of the unrepeatable character of his specific traits that underlie continuous advances in man's mind and humaneness. While rejecting the kind of anthropocentrism based on the conviction that since "the entire Universe revolves around man" and therefore he will never meet a premature end ... "unless the Universe itself should also collapse in accordance with its own predestination", modern scientific and philosophical thought more and more often comes to the idea of anthropocosmism, which views man as an organic and active part of the cosmos and of the Universe together with all its possible forms of life, intelligence and humaneness.[1]

This leads us to view the problem of the human race's immortality in a different perspective and to use the same approaches as when considering the relationship of the individual to the species, the part to the whole, and finiteness to infinity, finding "the infinite in the finite, the eternal in the transitory".[2]

While recognizing the possible finiteness of man's and of the human race's natural-biological existence, scientific philosophy not only asserts that in its transformations matter will repeatedly produce a thinking spirit "somewhere in another place and at another time",[3] but also its presence in infinitely diverse forms within the infinite Universe, as well as the probability that human intelligence itself will break through into the Universe. This can no longer be simply viewed as optimism in contrast to pessimistic ideas about the inevitable destruction of human reason and of its uniqueness that presumes our cosmic solitude. Rather it is scientific realism presupposing an infinite progress of reason and humaneness in the Universe. In particular, a discussion of the problem of finding extraterrestrial civilizations in the light of that humanistic aspect would make it possible to find new approaches to its understanding: one could assume the existence of forms of human intelligence that are higher than the human form but also of manifestations of humaneness that we currently do not understand and, hence, are deprived of possible contacts with them. There is a need, accordingly, for the new cosmic ethics to which Albert Schweitzer referred as an ethics of awe before life and its highest manifestations—intelligence and humaneness.

---

[1] Hoimar von Ditfurth, *Wir sind nicht nur von dieser Welt: Naturwissenschaft, Religion und die Zukunft des Menschen,* Hamburg, Hoffman und Campe, 1982, S. 137.

[2] Frederick Engels, *Dialectics of Nature,* p. 234.

[3] *Ibid.,* p. 39. It is true that modern cosmology speaks of a possible perishing of the Universe through either combustion or freezing. On the other hand, F. J. Dyson, for example, believes that through the law of adaptation "forms of life and of intelligence must lend themselves to transfer without losses from one environment to another, and this may ensure their infinity". (F. J. Dyson, "The Future of Will and the Future of Fate", *Nature,* No. 8, 1982, p. 70.)

But these are all speculations extending beyond the time parameters that define real and well-grounded forecasts concerning man's prospects in the context of currently existing social and natural-biological factors governing his existence and development. Nevertheless, it seems to me that within the framework of these time parameters, one may assert realistically that: a marvelous and infinite future awaits man and the human race provided that intelligence and humanism will prevail and will always continue to guide him. In man, Nature has attained the highest structural complexity; it has embodied in him a unique model that combines organically biological and social processes and material and spiritual ones. From a creation of Nature man has turned into its explorer and increasingly into its intelligent and skillful manager who is aware of his global and cosmic responsibility. The moral and ethical significance of this fact is truly immense, for it can stimulate a further awareness of the value of human life, and of man's intelligence and humaneness.

Man is an infinite Universe, and for man the Universe is an infinite "thread of life" running into eternity.

CONCLUSION

A NEW SYNTHESIS OF MAN, NATURE AND HUMANISM IN AN INTEGRAL CULTURE OF THE HUMAN RACE

A consideration of the problem of man's relation to science inevitably turns one's attention to culture, within which science is an important—but not the only—component of man's spiritual activity. It follows that the very possibility of a new synthesis of man and science is unavoidably associated with the development of an integral humanistic culture of mankind. It is therefore important to define what culture is taken to mean. Leaving aside the bourgeois conceptions, characteristically based on explicitly anti-humanistic approaches or on abstract neo-humanism, one cannot ignore the fact that different definitions of culture exist in Marxism. The general conception of culture adopted here links man's historically effective creative activity with the development of man as a subject of that activity.

That approach makes discernible many problems associated both with global processes of cultural progress and with tasks directly linked to the formation and development of the individual. In particular, it shows that while as a result of the multidirectional operation of many class-antagonistic factors, modern culture is fragmented, in the future it should acquire its integrity and unity. The disparity—which is growing under capitalism—between two types of culture, that of a ruling elite and that of cultural surrogates for the masses ("mass culture"), is unquestionably an acute social problem that must be solved, and is already being solved in socialist countries. Taken as such, however, that process is a reflection of some of culture's essential characteristics.

Today, culture's humanistic orientation is becoming more and more evident in the fact that culture is increasingly defined in terms of its relation to man and his development potential. At the same time, the integration of cultural development is promoted by the current internationalization of mankind's existence with the ensuing need to solve global problems. In short, both internal factors of cultural development, expressed in its increasing humanization, and socio-economic and other processes that are, on the whole, moving in the direction of internationalization, point to an increasing interaction among cultures and to their integration. This is not in conflict with the sovereignty and the distinctive originality of any one of them: in fact, the latter can even be promoted by the processes of interaction and integration under appropriate social conditions. It would appear that the strategy of cultural development must take into account the reciprocal influences of many processes and factors operating in different directions, and their integration into a single system. This goal may be fully achieved only by a new communist civilization whose essence corresponds to the humanistic essence of culture.

Such a conception of cultural development which emphasizes its humanistic and human factors is increasingly penetrating the consciousness of modern man, as was clearly evident, in particular, at the UNESCO Second World Conference on Cultural Policies, held in Mexico City in 1982. The materials prepared by UNESCO, the reports and contributions made by delegates from 129 countries and approximately 100 international organizations,[1] and the final documents (primarily the Mexico-city Declaration on cultural policies) all placed the emphasis on the humanistic approach to culture as the basic one, although its specific applications obviously differed depending on world views, ideological, political and social contexts.

The development of a common future humanistic culture depends, however, not only on social determinants but also on internal processes and tendencies associated with the specific characteristics of human activities in the spiritual sphere. Above all, this relates to the divergence and convergence of the "two cultures"—that of the sciences and that of the arts, which in the future must be in harmony with each other. This may be

[1] UNESCO *Courier,* August, 1982. The author of this book attended the Conference and read a paper entitled "Cultural Aspects of Development".

achieved through a close interaction of science and of the arts in the cognition and development of the rich potentials of human nature.

We have discussed the role that is played by science in the life of man, and we have seen that social factors and its own logic of development guide it towards the most intimate mysteries of human existence as it raises the most complicated problems of ethics and humanism, of value-oriented knowledge. Nevertheless, science is not all-powerful, or rather its power is not universal or absolute. In the course of its own development science stops short of solving certain mysteries of the human mind that lie beyond its reach. These problems may be solved, but their solution requires some other means, in particular the arts.

The reason lies in the profoundly individualized nature of works of art that results from their permanent orientation on man. A unique integral unity of man with the world in works of art distinguish art from science not only in terms of the means that are employed but also as regards their object which is always correlated with the personality of the artist and with his subjective perception of the world, while science seeks to move beyond these boundaries, guided by the principle of objectivity and nomotheticity. This is why science seeks to achieve a rigorous one-to-one correspondence in the perception of knowledge by man and finds corresponding means, its own language (using concepts, symbols, etc.), while works of art do not possess such an unambiguity; their perception, reflected in man's subjective world, produces a whole gamut of inherently individual shades that make that perception exceptionally diverse even though it is subordinated to a certain trend, to a general theme.

It is here that we find the secret of art's ability to influence man, his moral world, way of life, and behaviour. Science cannot fill man's soul to the full. In turning to art man exceeds the limits of rational unambiguity; art opens to him a world of mystery that does not lend itself to rational scientific cognition, and that is why man needs art as an organic part of what is contained within himself and in the world. In this connection, Niels Bohr observed: "The enrichment which art can give us originates in its power to remind us of harmonies beyond the grasp of systematic analysis."[1] One may also recall how deeply Darwin regretted towards the end of his life the loss of higher esthetic tastes, saying that if he were to relive his life he would seek to preserve the activity of those parts of the brain that are associated with esthetic tastes. "The loss of these tastes is a loss of happiness, and may possibly be injurious to the intellect, and more probably to the moral character, by enfeebling the emotional part of our nature."[2]

Because of art's orientation on man's moral and emotional world it is precisely in art that the moral-philosophical and humanistic problems of life and death, good and evil, human freedom and honour are raised in their most acute form. Like man himself, these problems, do not change as rapidly as scientific achievements do. In a certain sense, these problems are as "eternal" as is art which expresses them.

A social phenomenon, art possesses a multiplicity of functions: it performs an active socio-transforming function as a form of human activity; it plays an important heuristic role in cognition and the development of world views; it can anticipate many developments relating to man and the world; its educational and axiological functions are truly invaluable; finally, it plays a unique role as a source of esthetic pleasure (its hedonistic function). This has been vividly described by Einstein who said that personally he derived the greatest feeling of happiness from works of art and that there was no other sphere from which he derived such spiritual joy. Einstein added that he received more from Dostoyevsky than from any scientific thinker, and even more than from Gauss.[3]

One could cite many other statements made by great scientists concerning the immense importance that art has played in their lives and concerning the world of beauty and the unknown that it opens to man. Similarly, many great writers and painters

---

[1] Niels Bohr, *Atomic Physics and Human Knowledge,* John Wiley & Sons, Inc., New York, 1958, p. 79.

[2] *The Autobiography of Charles Darwin 1809-1882,* New York, Harcourt, Brace and Co., 1958, p. 139.

[3] A. Meshkovsky, *Albert Einshtein. Besedy s Einshteinom o teorii otnositelnosti i obshchei kartine mira* (Albert Einstein. Conversations with Einstein Concerning the Theory of Relativity and a General Image of the World), Moscow, 1922, p. 162.

have repeatedly stressed the enormous creative influence exerted on them by science, and by the beauty and greatness of the truth that originates in scientific research and discoveries. In particular, Gorky, referring to common elements of creativity in science and the arts, observed that "in both cases the main role is played by observations, comparison, a study of details; like a scientist, an artist must be imaginative and possess a gift of guessing—intuition."[1]

Science and the arts have common origins in man's labour, or productive activities, and therefore what they express in different forms is man's universal ability to create. In ancient times that ability was seen as unfragmented and integral in character. The subsequent course of development produced a sharp separation of science and the arts within the sphere of cognition. The former relied on logical rational forms and the latter, on intuitive irrational ones. The view that art could possess any cognitive functions was abandoned. Such extreme positions of scientism, which took a negative view of the potential of the arts, contributed to the development of various forms of anti-scientism that attributed an absolute role to intuitive and irrational acts of artistic or any other "enlightenment". This form of anti-scientism is still in wide use today in various forms of irrationalism, existentialist conceptions, and so on. A dominant role, however, is played by a scientistic tendency which places its hopes on the modern revolution in science and technology, allegedly capable of encompassing everything and bringing about the absolute triumph of scientific and technological rationality.

Yet conceptions suggesting that art must "be dissolved" in science and abandon its humanistic orientation are a threat not only to the development of art but also to that of science. For art also plays a major axiological and value-regulating role when expressed in the form of the *esthetics of scientific cognition,* a role similar to that played by sociology and the ethics of science.

Since forecasts to the effect that science would displace the

---

[1] *Gorky o nauke. Statyi, rechi, pisma, vospominaniya* (Gorky on Science. Articles, Speeches, Letters and Reminiscences), Moscow, 1964, p. 40.

arts have not materialized a different problem arose, namely, that of a separation of science and art. This process could lead, as some modern thinkers believe, to the formation of two independent cultures—a "scientific" and an "artistic" one. This raises a number of questions. Could the separation of science and art be a manifestation of the self-alienation and fragmentation of human personality? What if that separation has been exaggerated? Perhaps what at first appears to be a divergence is in fact an external manifestation of a new type of internal linkages between art and science?

Answers to these questions have been provided by Marxism which has demonstrated the heuristic effectiveness of its approach. It stresses that it is only by including both scientific and artistic activities into a common context of human activities, within which material production plays the role of a basis, that one can understand correctly their respective roles in the cognition and development of man. The development of culture which includes both scientific cognition and art is necessary to man's development as an active being. Marx noted that "an *objet d'art* creates a public that has artistic taste and is able to enjoy beauty—and the same can be said of any other product."[1]

· Today it has become clear that one-sided attempts to "sociologize" man's life and "make it scientific", as it were, have failed. Man is more complex than any of the limited scientistic conceptions of man that ever existed or, most probably, will ever exist. *Homo sapiens* is all woven of contradictions and earthly passions. And it is only as an earthly being that he affirms his own autonomous value and is of any interest on a cosmic plane.

Art seeks to express this unrepeatable uniqueness of man; it cultivates a creative element in everything—in perceiving truth, in transforming the world in accordance with the laws of the beautiful (Marx). Historically it is in art that the free creative individuality first developed. While having common features with science in the act of creativity itself, art, in its final creations, differs from the impersonal form in which the findings of

[1] Karl Marx, *A Contribution to the Critique of Political Economy,* p. 197.

scientific knowledge are recorded and bears the mark of the unique personality of its creator, thus permitting man to experience an involvement in the creative process. In short, art plays a specific role in the development of society's creative potential.

By disclosing the inner world of the individual, art brings us into communion with a certain personal and social ideal. In that sense, art is an exclusively human form of communion with the highest achievements of the human spirit. In experiencing works of art we enter into communion with the creative active element in a personality and are stimulated to follow its ideal.

Art reflects life in integral artistic images and is addressed to man as a whole rather than to any particular ability developed in a one-sided manner. This does not mean, of course, that as a result man can also be categorized in terms of some "totality" with regard to the development of his abilities including his capacity for scientific and artistic creative work. It would appear that even when they are all maximally developed various aspects of human nature will always express themselves in a different manner. And for all the integration, harmony and interaction between rational cognitive and emotional-image-wise artistic activities they will never merge completely. "Human diversity" which is based on the psychophysiological and biological-genetic differences among individuals constitutes a priceless capital that we should seek to multiply.

In short, instead of "merging", abilities for scientific and artistic work will flourish even more vividly and interact with increasing harmony. This refers especially to art which is often given an unjustifiably modest role in the context of human prospects. In fact the role of art in human life will continuously increase, as will its relevance in the development of the mankind's cultural values, including the ethical values that operate as regulators of scientific knowledge.

The value orientation of science and arts reflects some more general processes bearing on human culture as a whole and adding new colouring to the prospects of man and of the human race. While recognizing the historical progress, and consequently making a value judgement with regard to human history, we by no means interpret social development in providential terms nor attach an absolute value to current value orientations. "Men make their own history, but they do not make it just as they please; they do not make it under circumstances chosen by themselves, but under circumstances directly encountered, given and transmitted from the past."[1] Above all this means that the values that guide the actions of individuals are not something that is predefined but are developed by individuals. Secondly, while they may change historically these values are not randomly variable; they develop on the basis of the human race's past experience, and of the traditions of preceding generations. This confirms the enormous role of continuity in the material and spiritual development of society.

Such an approach helps avoid the extreme positions that often result from attempts to understand the respective roles of man's various value orientations, for example, his striving for truth and striving for beauty, between the assertion of good in the world and cultural tolerance, etc. One such extreme position is an attempt to develop an hierarchy of values ascending to some kind of ideal principle lying outside history. Another seeks to represent various value-oriented characteristics of the world as if they were completely unrelated and competed with one another in seeking to become a sole "categorical imperative" determining human behaviour. "Let the entire world collapse but truth (beauty, culture, etc.) must prevail"—such is a typical formula of value-oriented rigorism based on the exaggeration of only one of many essential values.

Both these extremes converge in one respect: both recognize the presence of a value-oriented characteristic dominating over the remaining determinants of human behaviour. This serves to dissolve the multicoloured world of spiritual values within the uniform light of some asserted absolute value. In stating that the realm of values is the historical outcome of man's own creativity we do not preclude the possibility that contradictions may arise between different value spheres governing human behaviour and in particular the possibility of a divergence and disharmony between the norms of ethics and those of culture. In that connection one should recognize both the danger of a moral fanaticism that may destroy culture and that of cultural all-permissiveness and omnivorousness. History, especially that of the

---

[1] Karl Marx, "The Eighteenth Brumaire of Louis Bonaparte", in Karl Marx, Frederick Engels, *Collected Works,* Vol. 11, 1979, p. 103.

twentieth century, gives quite a few examples of the perniciousness of attempts to view the world exclusively in terms of one value principle that is opposed to the entire remaining wealth of value orientation. But in rejecting the notion of a predefined harmony in man's system of value orientations we by no means reject the concept of such a harmony as an aim of spiritual creativity and as an ideal towards which individuals strive at each moment of their historical development.

While it appears to be difficult to be an "idealist" today, one cannot speak of a fundamental divergence between modern culture and ethics, for culture is ethical by its very nature. It is creative and engages in producing spiritual values of universal significance, thus opposing the power of evil, nihilistic and anti-cultural by definition. This does not mean, of course, that culture alone can overcome the evil that exists in the world, but it can be of considerable help in opposing it. But this requires that cultivated persons learn to recognize all possible modifications of evil, and recognize the strength that they can derive from their unity. We can find unity in understanding that the highest spiritual values are created by the entire human race and for the entire human race. Rather than create envy in some and false pride in one's exclusiveness in others the cultural achievements of individual countries should produce joy and a feeling of pride in man and his talent, and should unify persons rather than divide them. In this connection it is appropriate to recall Goethe's words concerning science and the arts: "As everything lofty and noble they belong to the entire world and they can be encouraged only by the mutual interaction of all contemporaries, with constant attention to what we have received from the past."[1]

As we have seen, some thinkers consider that scientific and technological progress is a factor which finds itself in opposition to humanistic culture and to the development of integral harmonious man. On this basis, wider generalizations have also been made concerning the allegedly insufficient resistance of humanistic consciousness, culture, and of the human elements in man vis-à-vis scientific knowledge and technical activities. Hu-

---

[1] Goethe, *Maximen und Reflexionen,* Aufbau-Verlag, Berlin und Weimar, 1982, p. 139.

manistic culture seems to be experiencing doubts as regards its mission, as in an age in which scientific and technical knowledge have achieved immense power, an ethical or esthetic orientation is often considered quixotic.

There is, however, a fundamentally different possibility for overcoming the conflict situation in which modern humanistic consciousness finds itself. Already in the nineteenth century Marx rejected as obviously futile the attempts to develop an inherently non-contradictory humanistic world view system and corresponding forms of practical behaviour. If the confliction arising in attempting to understand the humanistic principles of the self-development of the individual, culture and society is associated not with the personal merits or shortcomings of a thinker but with a spiritual-practical situation that he objectively confronts, then it is evident that the basic goal is to overcome and transform objectively existing structure of society.

This is why I believe that one should view the frequently encountered references to man's dehumanization and to a decline of culture, which essentially abandon the earlier humanistic cultural tradition of the past, as well as real practical efforts seeking to achieve mankind's socio-cultural and moral progress, not only as erroneous but also as profoundly immoral in view of the complicated world today, for, deliberately or not, their logic generates doubts concerning the possibility of man's further moral improvement.

In opposing the above conceptions, representatives of socialist culture consciously proclaim themselves to be the recipients of the lofty humanist legacy of the past and advocate real, communist humanism. Because it was at the cost of great suffering that mankind finally arrived at communist ideals, they contain a universal ethical meaning that stands in opposition to the views of the panic-prone prophets of "imminent self-destruction" of modern civilization and of its culture. Marxists associate the future of human civilization with the increasing humanization of culture. It is the development of a culture that places its pathos and ethos at the service of man and of his free and comprehensive development that can inspire and motivate action at the social and individual levels and the dialogues and concensus that it requires.

Throughout the twentieth century, and especially in the recent period, many voices have deplored the loss by man of his value as a human being, of his mutual interrelationship with his social environment and ultimately of the very meaning of his existence. That has been accompanied by gloomy expostulations concerning the advent of a spiritual crisis involving a crisis of science and the arts, a crisis of culture generally and even the demise of culture, and the approaching darkness of barbarism. It is true that the twentieth century has been a time of severe tests for humanistic culture. Still, man has kept the Prometheus' fire of creativity and preserved Hope, the leading virtue given to him by Prometheus and one of the most important and essentially irrational expressions of the soul's creative forces. Goethe noted that hope lives even at the foot of graves. It expresses man's boundless capacity to pursue ideals and aims that are like the Blue Bird that astonishes and beckons, producing strength and energy for creative action.

Creative work and hope give life to faith in the future, in historical progress, in man's power and invincibility. Hemingway rightly asserted that man was not created in order to be defeated and that man could be destroyed but not vanquished. Life is putting this proud conviction to tests that may be more severe than is necessary. Still, the continuing development of the wealth of human nature strengthens our hopes that reason and man's humaneness will triumph.

In making that conclusion we affirm the principle of new (real) humanism as the moral foundation of the scientific philosophy of man. I believe that it is in this principle that we find the logic of human history, the meaning of human existence and cognition, as well as man's hope for the future. It also contains the boundless spiritual power that will ensure a new synthesis and raise to new heights the integral humanistic culture of the human race, bringing it to a new and even more majestic Renaissance.

NAME INDEX

N

O

P

R

S

T

U

V

W

Y

Z